Thomas Fuller

The Cause and Cure of a Wounded Conscience

Also Triana, or, a Threefold Romanza, of Mariana, Paduana, and....

Thomas Fuller

The Cause and Cure of a Wounded Conscience
Also Triana, or, a Threefold Romanza, of Mariana, Paduana, and....

ISBN/EAN: 9783337008383

Printed in Europe, USA, Canada, Australia, Japan

Cover: Foto ©berggeist007 / pixelio.de

More available books at **www.hansebooks.com**

THE CAUSE AND CURE

OF

A WOUNDED CONSCIENCE;

ALSO

TRIANA; OR, A THREEFOLD ROMANZA,

OF

MARIANA, PADUANA, AND SABINA;
ORNITHOLOGIE, OR, THE SPEECH OF BIRDS;
AND ANTHEOLOGIA, OR, THE SPEECH OF FLOWERS.

BY

THOMAS FULLER, D.D.,

SOME TIME PREBENDARY OF OLD SARUM,
AND AUTHOR OF 'HISTORY OF THE WORTHIES OF ENGLAND,' 'ABEL REDIVIVUS,'
'THE CHURCH HISTORY OF BRITAIN,' ETC. ETC.

LONDON: WILLIAM TEGG.
1867.

TO THE RIGHT HONOURABLE
AND VIRTUOUS LADY
FRANCES MANNERS, COUNTESS OF RUTLAND.

MADAM,

By the judicial law of the Jews, if a servant* had children by a wife which was given him by his master, though he himself went forth free in the seventh year, yet his children did remain with his master, as the proper goods of his possession. I ever have been and shall be a servant to that noble family whence your *honour* is extracted. And of late in that house I have been wedded to the pleasant embraces of a private life, the fittest *wife* and meetest *helper* that can be provided for a student in troublesome times: and the same hath been bestowed upon me by the bounty of your noble brother, EDWARD LORD MONTAGUE.

* Exodus xxi. 4.

Wherefore what issue soever shall result from my mind, by his means most happily married to a retired life, must, of due, redound to his *honour*, as the sole proprietary of my pains during my present condition. Now this *book* is my eldest offspring, which had it been a son (I mean, had it been a work of masculine beauty and bigness), it should have waited as a *page* in dedication to his *honour*. But finding it to be of the weaker sex, little in strength and low in stature, may it be admitted (*madam*) to attend on your *ladyship*, his *honour's* sister.

I need not mind your *ladyship* how God hath measured outward happiness unto you by the *cubit of the sanctuary*, of the largest size, so that one would be perplexed to wish more than what your *ladyship* doth enjoy. My prayer to God shall be, that, shining as a pearl of grace here, you may shine as a star in glory hereafter. So resteth

Your Honour's,

In all Christian Offices,

THO. FULLER.

BOUGHTON,
January 25, 1646.

TO THE CHRISTIAN READER.

As one *was not anciently* to want a *wedding garment* at a marriage feast; so, now-a-days, wilfully to wear gaudy clothes at a funeral, is justly censurable as unsuiting with the occasion. Wherefore, in this sad subject, I have endeavoured to decline all light and luxurious expressions; and if I be found faulty therein, I cry and crave God and the reader pardon. Thus desiring that my pains may prove to the glory of God, thine, and my own edification, I rest,

Thine in Christ Jesus,
THO. FULLER.

THE CONTENTS OF THE SEVERAL DIALOGUES.

	PAGE
DIALOGUE I.—What a wounded conscience is, wherewith the godly and reprobate may be tortured	9
DIALOGUE II.—What use they are to make thereof, who neither hitherto were (nor haply hereafter shall be) visited with a wounded conscience	13
DIALOGUE III.—Three solemn seasons when men are surprised with wounded consciences	18
DIALOGUE IV.—The great torment of a wounded conscience proved by reasons and examples	23
DIALOGUE V.—Sovereign uses to be made of the torment of a wounded conscience	30
DIALOGUE VI.—That in some cases more repentance must be preached to a wounded conscience	34
DIALOGUE VII.—Only Christ is to be applied to souls truly contrite	39
DIALOGUE VIII.—Answers to the objections of a wounded conscience, drawn from the grievousness of his sins	44
DIALOGUE IX.—Answers to the objections of a wounded conscience, drawn from the slightness of his repentance	50
DIALOGUE X.—Answers to the objections of a wounded conscience, drawn from the feebleness of his faith	59
DIALOGUE XI.—God alone can satisfy all objections of a wounded conscience	62
DIALOGUE XII.—Means to be used by wounded consciences for the recovering of comfort	66

CONTENTS.

DIALOGUE XIII.—Four wholesome counsels for a wounded conscience to practise 76

DIALOGUE XIV.—Comfortable meditations for wounded consciences to muse upon 81

DIALOGUE XV.—That is not always the greatest sin whereof a man is guilty, wherewith his conscience is most pained for the present 87

DIALOGUE XVI.—Obstructions hindering the speedy flowing of comfort into a troubled soul 92

DIALOGUE XVII.—What is to be conceived of their final estate who die in a wounded conscience without any visible comfort 96

DIALOGUE XVIII.—Of the different time and manner of the coming of comfort to such who are healed of a wounded conscience 103

DIALOGUE XIX.—How such who are completely cured of a wounded conscience are to demean themselves . . 108

DIALOGUE XX.—Whether one cured of a wounded conscience be subject to a relapse 113

DIALOGUE XXI.—Whether it be lawful to pray for, or to pray against, or to praise God, for a wounded conscience. 117

THE CAUSE AND CURE

OF

A WOUNDED CONSCIENCE.

DIALOGUE I.

What a wounded Conscience is, wherewith the Godly and Reprobate may be tortured.

Timotheus. SEEING the best way never to know a *wounded conscience* by woeful experience, is speedily to know it by a sanctified consideration thereof: give me, I pray you, the description of a *wounded conscience*, in the highest degree thereof.

Philologus. It is a *conscience* frightened at the sight of sin,* and weight of God's wrath, even unto the despair of all pardon *during the present agony*.

Tim. Is there any difference betwixt a *broken*†

---

* Psalm xxxviii. 6.   † Psalm li. 17.

spirit and a *wounded* conscience, in this your acception?

Phil. Exceeding much : for a *broken spirit* is to be prayed and laboured for, as the most healthful and happy temper of the soul, letting in as much comfort as it leaks out sorrow for sin: whereas, a *wounded conscience* is a miserable malady of the mind, filling it for the present with despair.

Tim. In this your sense, is not the *conscience* wounded every time that the soul is smitten with guiltiness for any sin committed?

Phil. God forbid: otherwise his servants would be in a sad condition, as in the case of *David*,* smitten by his own heart, for being (as he thought) overbold with God's *anointed*, in cutting off the skirt of *Saul's* garment. Such hurts are presently healed by a *plaster* of *Christ's blood*, applied by *faith*, and never come to that height to be counted and called *wounded consciences*.

Tim. Are the *godly*, as well as the *wicked*, subject to the malady?

* 1 Sam. xxiv. 5.

Phil. Yes verily; *vessels of honour* as well as *vessels of wrath* in this world, are subject to the *knocks* and *bruises* of a *wounded conscience*. A patient *Job*, pious *David*, faithful *Paul*, may be vexed therewith, no less than a cursed *Cain*, perfidious *Achitophel*, or treacherous *Judas*.

Tim. What is the *difference* betwixt a *wounded conscience* in the *godly*, and in the *reprobate?*

Phil. None at all; oft times in the parties' apprehensions, both for the time being, conceiving their estates equally desperate; little, if any, in the wideness and anguish of the *wound* itself, which, for the time, may be as tedious and torturing in the *godly* as in the *wicked*.

Tim. How then do they differ?

Phil. Exceeding much in God's intention, gashing the *wicked*, as *malefactors*, out of justice, but lancing the *godly*, out of love, as a *surgeon* his *patients*. Likewise they differ in the issue and event of the *wound*, which ends in the eternal confusion of the one, but in the correction and amendment of the other.

Tim. Some have said, that in the midst of their pain, by this *mark* they may be distinguished; because the *godly*, when *wounded*, complain most of their *sins*, and the *wicked* of their *sufferings*.

Phil. I have heard as much; but dare not lay too much *stress* on this slender sign, (to make it generally true) for fear of failing. For *sorrow* for *sin*, and *sorrow* for *suffering*, are oft times so twisted and interwoven in the same person, yea in the same sigh and groan, that sometimes it is impossible for the party himself to separate and divide them in his own sense and feeling, as to know which proceeds from the one and which from the other. Only the all-seeing *eye* of an infinite God is able to discern and distinguish them.

Tim. Inform me concerning the nature of *wounded consciences* in the *wicked*.

Phil. Excuse me herein: I remember a *passage in S. Augustine,** who inquired what might be the cause that the *fall* of the *angels* is not plainly set down in the *Old Testament* with the manner

---

* *Angelicum vulnus verus medicus qualiter factum sit indicare noluit, dum illud postea curare non destinavit.* De Mirab. Scrip. lib. i. c. 2.

and circumstances thereof, resolves it thus: *God, like a wise surgeon, would not open that wound which he never intended to cure.* Of whose words, thus far I make use, that as it was not according to God's pleasure to restore the devils; so, it being above man's power to cure a *wounded conscience* in the *wicked*, I will not meddle with that which I cannot mend: only will insist on a wounded conscience in God's children, where, by God's blessing, one may be the instrument to give some ease and remedy unto their disease.

DIALOGUE II.

What use they are to make thereof, who neither hitherto were, nor haply hereafter shall be, visited with a wounded Conscience.

Tim. Are all God's children, either in their life or at their death, visited with a *wounded conscience?*

Phil. O no. God invites many with his *golden sceptre*, whom he never bruises with his *rod of iron*. Many, neither in their *conversion*, nor in the sequel of their *lives*, have ever felt that *pain* in such a manner and measure as amounts to a *wounded conscience*.

Tim. Must not the *pangs* in their *travail* of the *new birth* be painful unto them?

Phil. Painful, but in different degrees. The *Blessed Virgin Mary* (most hold) was delivered without any *pain*; as well may that *child* be *born* without *sorrow*, which is conceived without *sin*. The *women* of *Israel* were sprightful and lively, unlike the *Egyptians*.* The former favour, none can have, in their *spiritual travail*; the latter, some receive, who, though other whiles tasting of legal frights and fears, yet God so *preventeth*† *them with his blessings of goodness*, that they smart not so deeply therein as other men.

Tim. Who are those which commonly have such gentle usage in their *conversion*?

Phil. Generally such who never were notoriously profane, and have had the benefit of godly education from pious parents. In some corporations, the sons of freemen, bred under their fathers in their profession, may set up and exercise their

* Exod. i. 19. † Psalm xxi. 3.

father's trade, without ever being bound *apprentices thereunto*. Such children whose *parents* have been *citizens* of new *Jerusalem*,* and have been bred in the mystery of godliness, oftentimes are entered into religion without any *spirit of bondage* seizing upon them, a great benefit and rare blessing where God in his goodness is pleased to bestow it.

Tim. What may be the reason of God's dealing so differently with his own *servants*, that some of them are so deeply, and others not at all afflicted with *a wounded conscience?*

Phil. Even so, *Father*, because it *pleaseth thee*. Yet in *humility* these *reasons* may be assigned. 1. To show himself a *free agent*, not confined to follow the same precedent, and to deal with all as he doth with some. 2. To render the prospect of his proceedings the more pleasant to their sight who judiciously survey it, when they meet with so much diversity and variety therein. 3. That men being both ignorant when, and uncertain whether or not God will visit them with *wounded consciences*, may wait on him with humble hearts

* Gal. iv. 26; Eph. ii. 19; Heb. xii. 22.

in the work of their salvation, *looking as the eyes of the servants* to receive orders from the hand of their master ;* but what, when, and how, they know not, which quickens their daily expectations and diligent dependence on his pleasure.

Tim. I am one of those, whom God hitherto hath not humbled with a *wounded conscience :* give me some instructions for my behaviour.

Phil. First, be heartily thankful to God's infinite goodness, who hath not dealt thus with every one. Now because *repentance* hath two parts, *mourning* and *mending,* or *humiliation* and *reformation,* the more God hath abated thee in the former, out of his *gentleness,* the more must thou increase in the latter, out of thy *gratitude.* What thy *humiliation* hath wanted of other men, in the *depth* thereof, let thy *reformation* make up in the *breadth* thereof, spreading into an universal *obedience* unto all God's commandments. Well may he expect more *work* to be done by thy *hands,* who hath laid less *weight* to be borne on thy *shoulders.*

Tim. What other use must I make of God's kindness unto me ?

* Psalm cxxiii. 2.

Phil. You are bound the more patiently to bear all God's *rods, poverty, sickness, disgrace, captivity, &c.*, seeing God hath freed thee from the stinging scorpion of a *wounded conscience.*

Tim. How shall I demean myself for the time to come?

Phil. Be not high-minded, but fear; for thou canst not infallibly infer, that because thou hast not hitherto, hereafter thou shalt not, taste of a *wounded conscience.*

Tim. I will therefore for the future, with continual fear, wait for the coming thereof.

Phil. Wait not for it with servile fear, but watch against it with constant carefulness. There is a slavish fear to be visited with a *wounded conscience*, which fear is to be avoided, for it is opposite to the free spirit of grace, derogatory to the goodness of God in his gospel, destructive to spiritual joy, which we ought always to have, and dangerous to the soul, wrecking it with anxieties and unworthy suspicions. Thus to fear a *wounded conscience*, is in part to feel it, antedating one's misery and tormenting himself before the time, seeking for that he would be loth to find: like the wicked in the *Gospel*,* of whom it is said, *men's hearts failing them for fear, and looking for those things which are coming.* Far be such a *fear* from thee, and all good Christians.

Tim. What fear then is it, that you so lately recommended unto me?

Phil. One, consisting in the cautious avoiding of all causes and occasions of a *wounded conscience* conjoined with a confidence in God's goodness, that he will either preserve us from, or protect us in the torture thereof; and if he ever sends it, will sanctify it in us, to his glory, and our good. May I, you, and all God's servants, ever have this *noble fear* (as I may term it) in our hearts.

DIALOGUE III.

Three solemn Seasons when Men are surprised with wounded Consciences.

Tim. What are those times, wherein men most commonly are assaulted with *wounded consciences?*

Phil. So bad a *guest* may visit a man at any

* Luke xxi. 26.

hour of his life; for no season is unseasonable for God to be just, Satan to be mischievous, and sinful man to be miserable; yet it happens especially at three principal times.

Tim. Of these, which is the first?

Phil. In the twilight of a man's conversion, in the very conflict and combat betwixt nature and initial grace. For then he that formerly slept in carnal security, is awakened with his fearful condition. God, as he saith, *Psalm* l. 21, *setteth his sins in order before his eyes*. *Imprimis*, the sin of his conception. *Item*, the sins of his childhood. *Item*, of his youth. *Item*, of his man's estate, &c. Or, *Imprimis*, sins against the first table. *Item*, sins against the second; so many of ignorance, so many of knowledge, so many of presumption, severally sorted by themselves. He committed sins confusedly, huddling them up in heaps; but God *sets them in order*, and methodizes them to his hand.

Tim. Sins thus set in order must needs be a terrible sight.

Phil. Yes surely, the rather because the metaphor may seem taken from setting an *army in battle array.* At this conflict in his first conversion, *behold a troop* of sins *cometh*, and when God himself shall marshal them in *rank* and *file*, what guilty conscience is able to endure the furious charge of so great and well-ordered an army?

Tim. Suppose the party dies before he be completely converted, in this *twilight* condition as you term it, what then becomes of his soul, which may seem too good to dwell in outer darkness with devils, and too bad to go to the God of light?

Phil. Your supposition is impossible. Remember our discourse only concerns the godly. Now God never is father to abortive children, but to such who, according to his appointment, shall come to perfection.

Tim. Can they not therefore die in this *interim*, before the work of grace be wrought in them?

Phil. No verily. Christ's bones were in themselves breakable, but could not actually be broken by all the violence in the world, because God hath fore-decreed, *a bone of him shall not be broken.* So we confess God's children mortal; but all the power of devil or man may not, must not, shall not, cannot, kill them before their conversion, according to God's election of them to life, which must be fully accomplished.

Tim. What is the second solemn time wherein *wounded consciences* assault men?

Phil. After their conversion completed, and this either upon the committing of a conscience-wasting sin, such as *Tertullian* calls *peccatum devoratorium salutis,* or upon the undergoing of some heavy affliction of a bigger *standard* and proportion, blacker hue and complexion, than what befalls ordinary men, as in the case of *Job.*

Tim. Which is the third, and last time, when *wounded consciences* commonly walk abroad?

Phil. When men lie on their death-beds, Satan must now roar, or else for ever hold his peace: roar he may afterwards with very anger to vex himself, not with any hope to hurt us. There is mention in *Scripture of an evil day,* which is most applicable to the time of our *death.* We read also of an hour of temptation;* and the prophet † tells us there is a *moment, wherein God may seem to forsake us.* Now *Satan* being no less cunning to

* Rev. iii. 10.   † Isa. lviii. 7.

find out, than careful to make use of his time of advantage, in that *moment* of that *hour* of that *day*, will put hard for our souls, and we must expect a shrewd parting blow from him.

Tim. Your doleful prediction disheartens me, for fear I may be foiled in my last encounter.

Phil. Be of good comfort: through Christ we shall be victorious, both in dying and in death itself. Remember God's former favours bestowed upon thee. Indeed wicked men, from the premises of God's power, collect a conclusion of his weakness, *Psalm* lxxviii. 20. *Behold he smote the rock, that the waters gushed out, and the streams overflowed: can he give bread also? can he provide flesh for his people?* But God's children * by better *logic*, from the prepositions of God's former preservations, infer his power and pleasure to protect them for the future. Be assured, that *God*, who hath been the *God* of the *mountains*, and made our *mountains strong* in time of our *prosperity*, will also be the God of the valleys, and lead us safe *through the valley of the shadow of death.*†

* 1 Sam. xvii 36 ; 2 Cor. i. 10.   † Psalm xxxiii. 4.

DIALOGUE IV.

The great Torment of a wounded Conscience, proved by Reasons and Examples.

Tim. Is the pain of a *wounded conscience* so great as is pretended?

Phil. God * saith it, we have seen it, and others have felt it, whose complaints savour as little of dissimulation, as their cries in a fit of the cholic do of counterfeiting.

Tim. Whence comes this *wound* to be so great and grievous?

Phil. Six reasons may be assigned thereof. The first drawn from the *heaviness of the hand* which makes the *wound;* namely, God himself, conceived under the notion of an infinite angry judge. In all other afflictions, man encounters only with man, and in the worst temptations, only with *Satan;* but in a *wounded conscience,* he enters the lists immediately with God himself.

Tim. Whence is the second reason brought?

Phil. From the *sharpness* † of the *sword,* where-

* Prov. xviii. 14. † Heb. iv. 12.

with the *wound* is *made,* being the word of God, and the keen threatenings of the law therein contained. There is mention, *Gen.* iii. 24, of a *sword turning every way:* parallel whereto is the word of God in a *wounded conscience.* Man's heart is full of windings, turnings, and doublings, to shift and shun the stroke thereof if possible ; but this sword meets them wheresoever they move—it fetches and finds them out—it haunts and hunts them, forbidding them, during their agony, any entrance into the paradise of one comfortable thought.

Tim. Whence is the third reason derived?

Phil. From the *tenderness of the part* itself which is *wounded;* the *conscience* being one of the eyes of the soul, sensible of the smallest hurt. And when that *callum, schirrus,* or *incrustation,* drawn over it by nature, and hardened by custom in sin, is once flayed off, the *conscience* becomes so *pliant* and supple, that the least imaginable touch is painful unto it.

Tim. What is the fourth reason?

Phil. The *folly of the patient:* who being stung, hath not the wisdom to look up to Christ, the *brazen serpent,* but torments himself with his own activity. It was threatened to *Pashur*,* *I will make thee a terror to thyself*. So fares it with God's best saint during the fit of his perplexed *conscience;* he hears his own *voice*—he thinks, this is that which so often hath *sworn, lied,* talked *vainly, wantonly, wickedly;* his *voice* is a *terror to himself*. He sees his own *eyes* in a glass—he presently apprehends, these are those which shot forth so many envious, covetous, amorous *glances; his* eyes are a terror to himself. Sheep are observed to fly without cause, scared (as some say) with the *sound* of their own *feet*. Their feet knack, because they fly, and they fly, because their feet knack, an *emblem* of God's children in a *wounded conscience,* self-fearing, self-frightened.

Tim. What is the fifth reason which makes the pain so great?

Phil. Because *Satan* rakes his *claws* in the *reeking blood* of a *wounded conscience. Beelzebub*, the *devil's* name, signifies in *Hebrew the Lord of flies;* which excellently intimates his nature and employment: flies take their felicity about *sores* and *galled backs*, to infest and inflame them: So *Satan*

* Jer. xx. 4.

no sooner discovers (and that *bird* of *prey* hath *quick sight*) a soul *terror-struck*, but thither he hastes, and is busy to keep the *wound raw*—there he is in his throne to do mischief.

Tim. What is the sixth and last reason why a *wounded conscience* is so great a torment?

Phil. Because of the *impotency* and *invalidity of all earthly receipts* to give ease thereunto. For there is such a *gulf* of *disproportion* betwixt a mind-malady and body-medicines, that no carnal, corporal comforts can effectually work thereupon.

Tim. Yet wine in this case is prescribed in scripture. *Give wine to the heavy hearted, that they may remember their misery no more.**

Phil. Indeed if the *wound* be in the *spirits*, those *cursitors* betwixt soul and body, to recover their decay or consumption, *wine* may usefully be applied: but if the wound be in the *spirit*, in *scripture* phrase, all carnal, corporal comforts are utterly in vain.

Tim. Methinks *merry company* should do much to refresh him.

* Prov. xxxi. 6.

Phil. Alas! a man shall no longer be welcome in *merry company* than he is able to sing his part in their jovial concert. When a hunted *deer* runs for safeguard amongst the rest of the *herd,* they will not admit him into their company, but beat ~~heat~~ him off with their *horns,* out of principles of self-preservation, for fear the *hounds,* in pursuit of him, fall on them also. So hard it is for man or beast in misery, to find a faithful friend. In like manner, when a set of *bad-good-fellows* perceive one of their society dogged with God's *terrors* at his heels, they will forsake him as soon as they can, preferring his room, and declining his company, lest his sadness prove infectious to themselves. And now, if all six reasons be put together, *so heavy a hand,* smiting with *so sharp a sword* on so *tender a part* of so *foolish a patient,* whilst *Satan seeks to widen,* and *no worldly plaster can cure the wound,* it sufficiently proves a *wounded conscience* to be an exquisite torture.

Tim. Give me I pray an *example* hereof.

Phil. When *Adam* had eaten the *forbidden fruit* he tarried a time in *Paradise,* but took no contentment therein. The *sun* did shine as *bright,* the *rivers* ran as *clear* as ever before, *birds* sang as *sweetly*, *beasts* played as *pleasantly*, *flowers* smelled as *fragrant*, *herbs* grew as *fresh*, *fruits* flourished as *fair*, no *punctilio* of *pleasure* was either altered or abated. The *objects* were the same, but *Adam's eyes* were otherwise, his *nakedness stood in his light;* a *thorn* of *guiltiness grew in his heart* before any *thistles* sprang out of the ground; which made him not to seek for the *fairest fruits* to fill his *hunger*, but the *biggest leaves* to cover his *nakedness*. Thus a *wounded conscience* is able to unparadise *Paradise* itself.

Tim. Give me another instance.

Phil. CHRIST JESUS our *Saviour*: he was blinded, buffeted, scourged, scoffed at, had his hands and feet nailed on the *cross*, and all this while said nothing. But no sooner apprehended he his *Father* deserting him, groaning under the burthen of the sins of mankind imputed unto him, but presently the *Lamb* (who hitherto was *dumb before his shearer, and opened not his mouth*) for pain began to bleat, *My God, my God, why hast thou forsaken me?*

Tim. Why is a *wounded conscience* by *David*

resembled to *arrows: thine arrows stick fast in me?**

Phil. Because an arrow, especially if barbed, rakes and rends the flesh the more, the more metal the *wounded* party hath to strive and struggle with it: and a *guilty conscience* pierces the deeper, whilst a stout stomach with might and main seeks to outwrestle it.

Tim. May not a *wounded conscience* also work on the body to hasten and heighten the sickness thereof?

Phil. Yes verily, so that there may be employment for *Luke, the beloved physician,** (if the same person with the *Evangelist*) to exercise both his professions: but we meddle only with the malady of the mind, abstracted from any bodily indisposition.

* Psalm xxxviii. 2.   † Col. iv. 14.

DIALOGUE V.

Sovereign Uses to be made of the Torment of a wounded Conscience.

Tim. Seeing the torture of a *wounded conscience* is so great, what use is to be made thereof?

Phil. Very much. And first, it may make men sensible of the intolerable pain in *hell fire*. If the *mouth* of the *fiery furnace* into which the *children* were cast, was so hot that it burnt those which approached it, how hot was the *furnace* itself? If a *wounded conscience*, the *suburbs* of *hell*, be so painful, oh how extreme is that place, *where the worm never dieth, and the fire is never quenched?*

Tim. Did our *roaring boys* (as they call them) but seriously consider this, they would not wish GOD DAMN THEM, and GOD CONFOUND THEM so frequently as they do.

Phil. No verily. I read in *Theodoret* of the ancient *Donatists*, that they were so ambitious of martyrdom, (as they accounted it) that many of them meeting with a young *gentleman* requested of him that he would be pleased to kill them.

He, to confute their folly, condescended to their desire, on condition, that first they would submit to be fast bound: which being done, he gave order that they should be severely scourged, and then saved their lives. In application: When I hear such *riotous youths* wish that God would *damn* or *confound* them, I hope God will be more merciful than to *take them at their words*, and to grant them their wish; only I heartily desire that he would be pleased sharply to scourge them, and soundly to lash them with the frights and terrors of a *wounded conscience*. And I doubt not, but that they would so ill like the pain thereof, that they would revoke their wishes, as having little list, and less delight to taste of *hell* hereafter.

Tim. What other use is to be made of the pain of a *wounded conscience?*

Phil. To teach us seasonably to prevent what we cannot possibly endure. Let us shun the smallest sin, lest if we slight and neglect it, it by degrees fester and gangrene into a *wounded conscience*. One of the bravest *spirits* * that ever

---

* Sir Thomas *Norris*, president of *Munster, ex levi vulnere neglecto sublatus*. Camden's Elizab. An. 1641.

England bred, or *Ireland* buried, lost his life by a slight *hurt* neglected, as if it had been beneath his high mind to stoop to the dressing thereof, till it was too late. Let us take heed the stoutest of us be not so served in our souls. If we repent not presently of our sins committed, but carelessly contemn them, a *scratch* may quickly prove an *ulcer*; the rather, because the *flesh* of our mind, if I may so use the metaphor, is hard to heal, full of choleric and corrupt humours, and very ready to rankle.

Tim. What else may we gather for our instruction from the torture of a troubled mind?

Phil. To confute their cruelty, who out of sport or spite, willingly and wittingly wound *weak consciences*; like those uncharitable *Corinthians*,* who so far improve their liberty in things indifferent, as thereby to wound the *consciences* of their weak brethren.

Tim. Are not those ministers to blame, who, mistaking their *message*, instead of bringing the *gospel* of *peace*, frighten people with legal terrors into despair?

* 1 Cor. viii. 12.

Phil. I cannot commend their discretion, yet will not condemn their intention herein. No doubt their desire and design is pious, though they err in the pursuit and prosecution thereof, casting down them whom they cannot raise, and conjuring up the *spirit of bondage* which they cannot allay again: wherefore it is our wisest way, to interweave promises with threatenings, and not to leave open a pit of despair, but to cover it again with comfort.

Tim. Remaineth there not as yet another use of this point?

Phil. Yes, to teach us to pity and pray for those that have *afflicted consciences,* not like the wicked, *who persecute those whom God hath smitten, and talk to the grief of such whom he hath wounded.**

Tim. Yet *Eli* was a good man, who notwithstanding censured *Hannah,*† a woman of a sorrowful spirit, to be *drunk* with *wine.*

Phil. Imitate not *Eli* in committing, but amending his fault. Indeed his dim eyes could see *drunkenness* in *Hannah* where it was not, and

---

* Psalm lxix. 26. † 1 Sam. i. 13, 14.

could not see *sacrilege* and *adultery* in his own sons, where they were. Thus those who are most indulgent to their own, are most censorious of other's sins. But *Eli* afterwards perceiving his error, turned the condemning of *Hannah* into praying for her. In like manner, if in our passion we have prejudiced, or injured any *wounded consciences*, in cold blood let us make them the best amends and reparation.

DIALOGUE VI.

That in some Cases more Repentance must be preached to a wounded Conscience.

Tim. So much for the malady, now for the remedy. Suppose you come to a *wounded conscience*, what counsel will you prescribe him?

Phil. If after hearty prayer to God for his direction, he appeareth unto me, as yet, not truly penitent, in the first place I will press a deeper degree of repentance upon him.

Tim. O miserable comforter! more sorrow still! Take heed your eyes be not put out with that *smoking flax* you seek to quench, and your fingers wounded with the splinters of that *bruised reed* you go about to break.

Phil. Understand me, Sir. Better were my tongue spit out of my mouth, than to utter a word of grief to drive them to despair, who are truly contrite. But on the other side, I shall betray my trust, and be found an unfaithful dispenser of *divine mysteries,* to apply comfort to him who is not ripe and ready for it.

Tim. What harm would it do?

Phil. Raise him for the present, and ruin him, without God's greater mercy, for the future. For comfort daubed on, on a foul soul, will not stick long upon it: and instead of pouring in, I shall spill the precious oil of God's mercy. Yea, I may justly bring a *wounded conscience* upon myself, for dealing deceitful in my *stewardship.*

Tim. Is it possible one may not be soundly humbled, and yet have a *wounded conscience.*

Phil. Most possible; for a *wounded conscience* is often inflicted as a punishment for lack of true repentance: great is the difference betwixt a man's being frightened at, and humbled for, his sins. One may passively be cast down by God's terrors, and yet not willingly throw himself down as he ought at God's footstool.

Tim. Seeing his pain is so pitiful as you have formerly proved; why would you add more grief unto him?

Phil. I would not add grief to him, but alter grief in him; making his sorrow, not greater, but better. I would endeavour to change his dismal, doleful dejection, his hideous and horrible heaviness, his bitter exclamations, which seem to me much mixed in him, with pride, impatience, and impenitence, into a willing submission to God's pleasure, and into a kindly, gentle, tender, gospel repentance, for his sins.

Tim. But there are some now-a-days who maintain that a child of God, after his first conversion, needs not any new repentance for sin all the days of his life.

Phil. They defend a grievous and dangerous error. Consider what two petitions Christ couples together in his *prayer*. When my body, which every day is hungry, can live without God's *giving it daily bread*, then and no sooner shall I believe that my soul, which daily sinneth, can spiritually live, without God's *forgiving it its trespasses.*

Tim. But such allege, in proof of their opinion, that a man hath his person justified before God, not by pieces and parcels, but at once and for ever in his conversion.

Phil. This being granted doth not favour their error. We confess God finished the creation of the world, and all therein in six days, and then rested from that work; yet so, that his daily preserving of all things by his providence, may still be accounted a constant and continued creation. We acknowledge in like manner, a child of God justified at once in his conversion, when he is fully and freely estated in God's favour. And yet, seeing every daily sin by him committed, is an aversion from God, and his daily repentance a conversion to God, his justification in this respect may be conceived entirely continued all the days of his life.

Tim. What is the difference betwixt the first repentance, and this renewed repentance?

Phil. The former is as it were the putting of life into a dead man, the latter, the recovering of a sick man from a dangerous wound : by the former, sight to the blind is simply restored, and eyes given him; in the latter, only a film is removed, drawn over the eyes, and hindering their actual sight. By the first, we have a right title to the kingdom of Heaven : by our second repentance, we have a new claim to Heaven, by virtue of our old title. Thus these two kinds of repentance may be differenced and distinguished, though otherwise they meet and agree in general qualities : both having sin for their cause, sorrow for their companion, and pardon for their consequent and effect.

Tim. But do not God's children after committing grievous sins, and before their renewing their repentance, remain still heirs of Heaven, married to Christ, and citizens of the new Jerusalem ?

Phil. Heirs of Heaven they are, but disinheritable for their misdemeanour. Married still to Christ, but deserving to be divorced for their adulteries. Citizens of Heaven, but yet outlawed, so that they can recover no right, and receive no benefit, till their outlawry be reversed.

Tim. Where doth God in Scripture enjoin this second repentance on his own children?

Phil. In several places. He threatens the church of *Ephesus* (the best of the seven) with *removing the candlestick from them, except they repent :** and Christ tells his own disciples, true converts before, but then guilty of ambitious thoughts, that *except ye be converted ye shall not enter into the kingdom of Heaven.*† Here is conversion after conversion, being a solemn turning from some particular sin; in relation to which it is not absurd to say, that there is justification after justification; the latter as following in time, so flowing from the former.

DIALOGUE VII.

Only Christ is to be applied to Souls truly contrite.

Tim. But suppose the person in the minister's apprehension heartily humbled for sin, what then is to be done?

Phil. No corrosives, all cordials; no vinegar

* Rev. ii. 5.   † Mat. xviii. 3.

all oil; no law, all gospel must be presented unto him. Here, blessed the lips, yea beautiful the feet of him that bringeth the tidings of *peace.* As *Elisha,** when reviving the *son of the Shunamite,* laid his *mouth to the mouth of the child,* so the gaping orifice of Christ's *wounds* must spiritually, by preaching, be put close to the *mouth* of the *wounds* of a conscience: happy that skilful *architect* that can show the sick man that the *head stone* † of his spiritual building must be laid with shouts, crying, *grace, grace.*

Tim. Which do you count the *head stone* of the building, that which is first or last laid?

Phil. The foundation is the *head stone* in honour, the *top stone* is the *head stone* in height. The former the *head stone* in strength, the latter in stature. It seemeth that God's spirit, of set purpose made use of a doubtful word, to show that the whole fabric of our salvation, whether as founded, or as finished, is the only work of God's grace alone. Christ is the *alpha* and *omega* thereof, not excluding all the letters in the alphabet interposed.

* 2 Kings iv. 34.   † Zech. iv. 7.

Tim. How must the minister preach Christ to an *afflicted conscience?*

Phil. He must crucify him before his eyes, lively setting him forth; naked, to clothe him; wounded, to cure him; dying, to save him. He is to expound and explain unto him, the dignity of his person, preciousness of his blood, plenteousness of his mercy, in all those loving relations wherein the scripture presents him: a kind *father* to a prodigal *child,* a careful *hen* to a scattered *chicken,* a good *shepherd* that bringeth his lost *sheep* back on his *shoulders.*

Tim. Spare me one question, why doth he not drive the sheep before him, especially seeing it was lively enough to lose itself.—

Phil. First, because though it had wildness too much to go astray, it had not wisdom enough to go right. Secondly, because probably the silly sheep had tired itself with wandring; *Habakuk* ii. 13, " *The people shall weary themselves for very vanity,*" and therefore the kind shepherd brings it home on his own shoulders.

Tim. Pardon my interruption, and proceed, how Christ is to be held forth.

Phil. The latitude and extent of his love, his invitation without exception, are powerfully to be pressed; *every one that thirsteth, all ye that are heavy laden, whosoever believeth,* and the many promises of mercy are effectually to be tendered unto him.

Tim. Where are those promises in *scripture?*

Phil. Or rather, where are they not? for they are harder to be missed than to be met with. Open the *Bible* (as he who drew his bow in battle*) at a venture. If thou lightest on an *historical place,* behold precedents; if on a doctrinal, promises of comfort. For the latter, observe these particulars; *Gen.* iii. 15, *Exo.* xxxiii. 6, *Isa.* xl. 1, *Isa.* liv. 11, *Mat.* xi. 28, *Mat.* xii. 20, 1 *Cor.* x. 13, *Heb.* xiii. 5, &c.

Tim. Are these more principal places of consolation than any other in the *Bible?*

Phil. I know there is no choosing, where all things are choicest. Whosoever shall select some pearls out of such a *heap,* shall leave behind as precious as any he takes, both in his own and others' judgment: yea, which is more, the same

* 1 Kings xxii. 34.

man at several times may in his apprehension prefer several promises as best, formerly most affected with one place, for the present more delighted with another: and afterwards conceiving comfort therein not so clear, choose other places as more pregnant and pertinent to his purpose. Thus God orders it, that divers men (and perchance the same man at different times) make use of all his promises, gleaning and gathering comfort, not only in one furrow, land, or furlong, but as it is scattered clean through the whole *field* of the *scripture*.

Tim. Must ministers have variety of several comfortable promises?

Phil. Yes, surely: such *masters of the assembly* being to enter and fasten consolation in an afflicted soul, need have many *nails* provided beforehand, that if some, for the present, chance to drive untowardly, as splitting, going awry, turned crooked or blunt, they may have others in the room thereof.

Tim. But grant Christ held out never so plainly, pressed never so powerfully, yet all is in vain, except God inwardly with his spirit persuade the wounded conscience to believe the truth of what he saith.

Phil. This is an undoubted truth, for one may lay the *bread of life* on their trencher, and cannot force them to feed on it. One may bring them down to the *spring of life,* but cannot make them drink of the waters thereof: and therefore in the cure of a *wounded conscience,* God is all in all, only the touch of his hand can heal this *king's evil, I kill and make alive, I wound and I heal, neither is there any that can deliver out of my hand.**

DIALOGUE VIII.

Answers to the objections of a wounded Conscience drawn from the grievousness of his sins.

Tim. Give me leave now, Sir, to personate and represent a *wounded conscience,* and to allege and enforce such principal objection wherewith generally they are grieved.

Phil. With all my heart, and God bless my endeavours in answering them.

* Deut. xxxii. 39.

Tim. But first I would be satisfied how it comes to pass that men in a *wounded conscience* have their parts so presently improved. The *Jews* did question concerning our *Saviour, How knoweth this man letters being never learned.** But here the doubt and difficulty is greater. How come *simple people* so subtle on a sudden, to oppose with that advantage and vehemence, that it would puzzle a good and grave *divine* to answer them.

Phil. Two reasons may be rendered thereof. 1. Because a man in a distemper is stronger than when he is in his perfect health. What *Samson's* are some in the fit of a *fever?* Then their spirits, being raised by the violence of their disease, push with all their power. So is it in the agony of a distressed soul, every string thereof is strained to the height, and a man becomes more than himself to object against himself in a fit of despair.

Tim. What is the other reason?

Phil. Satan himself, that subtle *sophister*, assists them. He forms their arguments, frames their objections, fits their distinctions, shapes their evasions; and this *discomforter* (aping God's spirit,

* John vii. 15.

the comforter, *John* xiv. 26) *bringeth all things to their remembrance* which they have heard or read, to dishearten them. Need therefore have ministers, when they meddle with afflicted men, to call to Heaven aforehand to assist them, being sure they shall have hell itself to oppose them.

Tim. To come now to the objections which *afflicted consciences* commonly make: they may be reduced to three principal heads; either drawn from the greatness and grievousness of their sins, or from the slightness and lightness of their repentance, or from the faintness and feebleness of their faith; I begin with the objections of the first form.

Phil. I approve your method, pray proceed.

Tim. First, Sir, even since my conversion, I have been guilty of many grievous sins; and, which is worse, of the same sin many times committed. Happy *Judah*, who though once committing incest with *Thamar*, yet the text saith, that afterwards he *knew her again no more.** But I, vile wretch, have often refallen into the same offence.

* Gen. xxxviii. 36.

Phil. All this is answered in God's *promise* in the *prophet,—Though your sins be as scarlet, I will make them as snow.** Consider how the *Tyrian scarlet* was dyed, not superficially dipped, but thoroughly drenched in the liquor that coloured it, as thy soul in custom of sinning. Then was it taken out for a time and dried, put in again, soaked and sodden the second time in the fat; called therefore δίβαφον, twice dyed; as thou complainest thou hast been by relapsing into the same sin. Yea, the colour so incorporated into the cloth not drawn over, but diving into the very heart of the wool, that rub a scarlet rag on what is white, and it will bestow a reddish tincture upon it; as perchance, thy sinful practice and precedent have also infected those which were formerly good, by thy badness. Yet such scarlet sins so solemnly and substantially coloured, are easily washed white in the blood of our *Saviour.*

Tim. But, Sir, I have sinned against most serious resolutions, yea against most solemn vows, which I have made to the contrary.

Phil. Vow-breaking, though a grievous sin, is

* Isaiah i. 18.

pardonable on unfeigned repentance. If thou hast broken a vow, tie a *knot* on it to make it hold together again. It is spiritual thrift, and no misbecoming baseness, to piece and join thy neglected promises with fresh ones. So shall thy vow in effect be not broken when new mended: and remain the same, though not by one entire continuation, yet by a constant successive renovation thereof. Thus *Jacob* renewed his neglected vow of going to *Bethel*;* and this must thou do, reinforce thy broken vows, if of *moment and material*.

Tim. What mean you by the addition of that clause, if *of moment and material?*

Phil. To deal plainly, I dislike many vows men make, as of reading just so much, and praying so often every day, of confining themselves to such a strict proportion of meat, drink, sleep, recreation, &c. Many things may be well done, which are ill vowed. Such particular vows men must be very sparing how they make. First, because they savour somewhat of will-worship. Secondly, small glory accrues to God thereby. Thirdly, the dignity of vows is disgraced by descending to too

---
* Compare Gen. xxviii. 20, with Gen. xxxv. 1.

trivial particulars. Fourthly, Satan hath ground given him to throw at us with a more steady aim. Lastly, such vows, instead of being cords to tie us faster to God, prove knots to entangle our consciences: hard to be kept, but oh! how heavy when broken! Wherefore setting such vows aside, let us be careful, with *David*, to keep that grand and general vow; *I have sworn, and I will perform it, that I will keep thy righteous judgments.**

Tim. But, Sir, I have committed the *sin* against the *Holy Ghost*, which the *Saviour* of mankind pronounceth unpardonable, and therefore all your counsels and comforts unto me are in vain.

Phil. The devil, the father of lies, hath added this lie to those which he hath told before, in persuading thee thou hast committed the *sin* against the *Holy Ghost*. For that sin is ever attended with these two symptoms. First, the party guilty thereof never grieves for it, nor conceives the least sorrow in his heart for the sin he hath committed. The second, which followeth on the former, he never wishes or desires any pardon,

* Psal. cxix. 106.

but is delighted and pleased with his present condition. Now, if thou canst truly say that thy sins are a burden unto thee, that thou dost desire forgiveness, and wouldest give any thing to compass and obtain it, be of good comfort, thou hast not as yet, and, by God's grace, never shalt commit that unpardonable offence. I will not define how near thou hast been unto it. As *David* said to *Jonathan, there is not a hair's breadth betwixt death and me :* so it may be thou hast missed it very narrowly, but assure thyself thou art not as yet guilty thereof.

DIALOGUE IX.

Answers to the objections of a wounded Conscience drawn from the slightness of his Repentance.

Tim. I believe my sins are pardonable in themselves, but alas, my *stony heart* is such, that it cannot relent and repent, and therefore no hope of my salvation.

Phil. Wouldest thou sincerely repent? thou dost repent. The women that came to embalm

Christ, did carefully forecast with themselves, *who shall roll away the stone from the door of the sepulchre?* * Alas, their frail, faint, feeble arms, were unable to remove such a weight. But what follows? *And when they looked, they saw that the stone was rolled away, for it was very great.* In like manner, when a soul is truly troubled about the mighty burden of his stony heart interposed, hindering him from coming to Christ; I say when he is seriously and sincerely solicitous about that impediment, such desiring is a doing, such wishing is a working. Do thou but take care it may be removed, and God will take order it shall be removed.

Tim. But, Sir, I cannot weep for my sins; my eyes are like the pit wherein *Joseph* was put; there is no water in them, I cannot squeeze one tear out of them.

Phil. Before I come to answer your objection, I must premise a profitable observation. I have taken notice of a strange opposition betwixt the tongues and eyes of such as have *troubled consciences.* Their tongues some have known (and I

* Mark xvi. 3.

have heard) complain that they cannot weep for their sins, when at that instant their eyes have plentifully shed store of tears: not that they spake out of dissimulation, but distraction. So sometimes have I smiled at the simplicity of a child, who being amazed, and demanded whether or no he could speak? hath answered, *no*. If in like manner at the sight of such a contradiction betwixt the words and deeds of one in the agony of a *wounded conscience*, we should chance to smile, knew us not to jeer, but joy, perceiving the party in a better condition than he conceiveth himself.

Tim. This your observation may be comfortable to others, but is impertinent to me. For as I told you, I have by nature such dry eyes that they will afford no moisture to bemoan my sins.

Phil. Then it is a natural defect, and no moral default, so by consequence a suffering and no sin which God will punish. God doth not expect the pipe should run water, where he put none into the cistern. Know also, their hearts may be fountains whose eyes are flints, and may inwardly bleed who do not outwardly weep. Besides, Christ was sent to *preach comfort*, not to such only as *weep*, but *mourn* in *Zion.** Yea, if thou canst squeeze out no liquor, offer to God the empty bottles; instead of tears, tender and present thine eyes unto him. And though thou are water-bound, be not wind-bound also, sigh where thou canst not sob, and let thy *lungs* do what thine *eyes* cannot perform.

Tim. You say something though I cannot weep, in case I could soundly sorrow for my sins. But alas, for temporal losses and crosses, I am like *Rachael, lamenting for her children, and would not be comforted.* But my sorrow for my sins is so small that it appears none at all in proportion.

Phil. In the best saints of God, their sorrow for their sins being measured with the sorrow for their sufferings, in one respect, will fall short of it, in another must equal it, and in a third respect doth exceed and go beyond it. Sorrow for sins falls short of sorrow for sufferings, in loud lamenting or violent uttering itself in outward expressions thereof; as in roaring, wringing the hands, rending the hair, and the like. Secondly, both sorrows are equal in their truth and sincerity, both far from

* Isaiah li. 3.

hypocrisy, free from dissimulation, really hearty, cordial, uncounterfeited. Lastly, sorrow for sin exceeds sorrow for suffering, in the continuance and durableness thereof: the other, like a land-flood, quickly come, quickly gone; this is a continual dropping or running river, keeping a constant stream. *My sins*, saith *David, are ever before me;* so also is the sorrow for sin in the soul of a child of God, morning, evening, day, night, when sick, when sound, feasting, fasting, at home, abroad, ever within him. This grief begins at his conversion, continues all his life, ends only at his death.

Tim. Proceed I pray in this comfortable point.

Phil. It may still be made plainer by comparing two diseases together, the *toothache* and *consumption*. Such as are troubled with the former, shriek and cry out, troublesome to themselves, and others in the same and next roof: and no wonder, the *mouth* itself being *plaintiff*, if setting forth its own grievances to the full. Yet the *toothache* is known to be no mortal malady, having kept some from their beds, seldom sent them to their graves; hindered the sleep of many, hastened the death of few. On the other side, he that hath an incurable *consumption* saith little, cries less, but grieves most of all. Alas, he must be a good husband of the little breath left in his broken lungs, not to spend it in sighing, but in living, he makes no noise, is quiet and silent; yet none will say but that his inward grief is greater than the former.

Tim. How apply you this comparison to my objection?

Phil. In corporal calamities, thou complainest more, like him in the *toothache*, but thy sorrow for thy sin, like a *consumption*, which lies at thy heart, hath more solid heaviness therein. Thou dost take in more grief for thy sins, though thou mayest take on more grievously for thy sufferings.

Tim. This were something if my sorrow for sin were sincere, but alas, I am but a hypocrite. There is mention in the prophet of God's *besom of destruction;** now the trust of a hypocrite, *Job* viii. 14, is called a *spider's web*, here is my case, when God's *besom* meets with the *cobwebs* of my hypocrisy, I shall be *swept* into *hell fire*.

* Isaiah xiv. 23.

Phil. I answer, first in *general:* I am glad to hear this objection come from thee, for self-suspicion of hypocrisy is a hopeful symptom of sincerity. It is a *David* that cries out, *as for me I am poor and needy;* but lukewarm *Laodicea* that brags, *I am rich and want nothing.*

Tim. Answer I pray the objection in *particular.*

Phil. Presently, when I have premised the great difference, betwixt a man's being a *hypocrite,* and having some hypocrisy in him. Wicked men are like the *apples* of *Sodom,** seemingly fair, but nothing but *ashes* within. The best of God's servants are like sound *apples,* lying in a dusty loft (living in a wicked world), gathering much dust about them, so that they must be rubbed or pared before they can be eaten. Such notwithstanding are sincere, and by the following *marks* may examine themselves.

Tim. But some in the present day are utter enemies to all *marks* of sincerity, *counting* it needless for *preachers* to propound, or *people* to apply them.

Phil. I know as much; but it is the worst sign,

* *Solinus Polyhistor* in *Judea.*

when men of this description hate all signs: but no wonder if the *foundered horse* cannot abide the *smith's pincers.*

Tim. Proceed I pray in your signs of sincerity.

Phil. Art thou careful to order thy very *thoughts*, because the infinite *searcher of the heart* doth behold them? Dost thou freely and fully confess thy sins to God, spreading them open in his presence, without any desire or endeavour to deny, dissemble, defend, excuse, or extenuate them? Dost thou delight in an universal *obedience* to all *God's laws*, not thinking, with the superstitious *Jews*, by over-keeping the fourth *commandment* to make reparation to God for breaking all the rest. Dost thou love their persons and preaching best, who most clearly discover thine own faults and corruptions unto thee? Dost thou strive against thy revengeful nature, not only to forgive those who have offended thee, but also to wait an occasion with humility to render a suitable favour to them? Dost thou love grace and goodness even in those who differ from thee in point of *opinion* and *civil controversies?* Canst thou be sorrowful for the sins of others no whit relating unto thee,

F

merely because the glory of a good God suffers by their profaneness?

Tim. Why do you make these to be the signs of sincerity?

Phil. Because there are but two principles which act in men's hearts, namely, *nature* and *grace* ; or, as *Christ* distinguishes them, *flesh* and *blood*, and *our father which is in Heaven*. Now seeing these actions, by us propounded, are either against or above nature, it doth necessarily follow, that where they are found they flow from saving grace. For what is higher than the roof and very pinnacle, as I may say, of nature, cannot be lower than the bottom and beginning of grace.

Tim. Perchance, on serious search I may make hard shift to find some one or two of these signs, but not all of them, in my heart.

Phil. As I will not bow to flatter any, so I will fall down as far as truth will give me leave, to reach comfort to the humble, to whom it is due. Know to thy further consolation, that where some of these signs truly are, there are more, yea all of them, though not so visible and conspicuous, but in a dimmer and darker degree. When we behold *violets* and *primroses* fairly to flourish, we conclude the dead of the winter is past, though as yet no *roses* or *July flowers* appear, which, long after, lie hid in their leaves or lurk in their roots, but in due time will discover themselves. If some of these signs be above ground in thy sight, others are under ground in thy heart; and though the former started first, the other will follow in order, it being plain that thou art passed from death unto life, by this hopeful and happy spring of some signs in thy heart.

DIALOGUE X.

Answers to the Objection of a wounded Conscience, drawn from the feebleness of his faith.

Tim. But faith is that which must apply Christ unto us; whilst (alas!) the hand of my faith hath not only the shaking, but the dead palsy, it can neither hold nor feel anything.

Phil. If thou canst not hold God, do but touch him and he shall hold thee, and put feeling into thee. Saint *Paul* saith,* *If that I may apprehend that, for which also I am apprehended of Christ Jesus.* It is not *Paul's* apprehending of *Christ*, but *Christ* apprehending of *Paul*, doth the deed.

Tim. But I am sure my faith is not sound, because it is not attended with assurance of salvation. For I doubt (not to say despair) thereof. Whereas Divines hold, that the Essence of saving faith consists in a certainty to be saved.

Phil. Such deliver both a false and dangerous doctrine; as the careless mother† killed her little *infant, for she overlaid it.* So this opinion would press many weak faiths to death, by laying a greater weight upon them than they can bear, or God doth impose; whereas to be assured of salvation, is not a part of every true faith, but only an effect of some strong faiths, and that also not always, but at some times.

Tim. Is not certainty of salvation a part of every true faith?

Phil. No verily; much less is it the life and formality of faith, which consists only in a recum-

* Phil. iii. 19. † 1 Kings iii. 19.

bency on God in Christ, with *Job's* resolution,* *Though he slay me, yet will I trust in him.* Such an adherence, without an assurance, is sufficient by God's mercy to save thy soul. Those that say that none have a sincere faith without a certainty of salvation, may with as much truth maintain, that none are the King's loyal Subjects but such as are his *Favourites.*

Tim. Is then *assurance of salvation* a peculiar personal *favour,* indulged by God only to some particular persons?

Phil. Yes verily. Though the *salvation* of all God's *servants* be sure in itself, yet it is only *assured* to the apprehensions of some *select* people, and that at some times. For it is too fine *fare* for the best man to *feed* on every day.

Tim. May they that have this *assurance* afterwards lose it?

Phil. Undoubtedly they may. God first is gracious to give it them, they for a time careful to keep it; then negligently lose it, then sorrowfully seek it. God again is bountiful to restore it, they happy to recover it; for a while diligent

---

* Job. xiii. 15.

to regain it, then again foolish to forfeit it; and so the same changes in one's lifetime, often over and over again.

Tim. But some will say, If I may be infallibly saved without this *Assurance,* I will never endeavour to attain it.

Phil. I would have *covered* my *flowers* if I had suspected such *spiders* would have sucked them. One may go to heaven without this *Assurance*, as certainly, but not so cheerfully; and therefore prudence to obtain our own comfort, and piety to obey God's command, obliges us all to *give diligence to make our calling and election sure,* both in itself and in our apprehension.

DIALOGUE XI.

God alone can satisfy all Objections of a wounded Conscience.

Tim. But, Sir, these your answers are no whit satisfactory unto me.

Phil. An answer may be satisfactory to the objection, both in itself and in the judgment of all unprejudiced hearers, and yet not satisfactory to the objector, and that in two cases. First, when he is possessed with the spirit of peevishness and perverseness. It is lost labour to seek to feed and fill those who have a greedy *horseleech* of cavilling in their heart, crying, *give, give.*

Tim. What is the second case?

Phil. When the bitterness of his soul is so great and grievous, that he is like the *Israelites* * in *Egypt,* who *hearkened not to Moses, for anguish of spirit and for cruel bondage.* Now, as those who have meat before them and will not eat, deserve to starve without pity; so such are much to be bemoaned, who through some impediment in their mouth, throat, or stomach, cannot chew, swallow, or digest, comfort presented unto them.

Tim, Such is my condition, what then is to be done unto me?

Phil. I must change my precepts to thee into prayers for thee, that God would *satisfy thee early with his mercy, that thou mayest rejoice.*† Ministers may endeavour it in vain; whilst they quell one scruple, they start another; whilst they fill one

* Exod. vi. 9. † Psalm xc.

corner of a wounded conscience with comfort, another is empty. Only God can so satisfy the soul, that each chink and cranny therein shall be filled with spiritual joy.

Tim. What is the difference betwixt God's and *man's speaking peace* to a *troubled spirit.*

Phil. Man can neither make him to whom he speaks, to hear what he says, or believe what he hears. God speaks with *authority*, and doth both. His words give *hearing* to the *deaf*, and *faith* to the *infidel.* When, not the *mother* of *Christ*, but *Christ himself*, shall salute a *sick soul* with *peace be unto thee*, it will leap for joy, as *John* the *babe* sprang, though imprisoned in the dark womb of his mother. Thus the *offender* is not comforted, though many of the spectators and under *officers* tell him he shall be *pardoned*, until he hears the same from the mouth of the Judge himself, who hath power and place to forgive him; and then his heart revives with *comfort.*

Tim. God send me such comfort: in the mean time I am thankful unto you for the answers you have given me.

Phil. All that I will add is this. The *Lace-*

demonians had a law, that if a bad man, or one disesteemed of the people, chanced to give good *counsel*, he was to stand by, and another, against whose person the people had no prejudice, was to speak over the same words which the former had uttered. I am most sensible to myself of my own wickedness, and how justly I am subject to exception. Only my prayer shall be, that whilst I stand by, and am silent, God's spirit, which is free from any fault, and full of all perfection, would be pleased to repeat in thy heart the self-same answers I have given to your objections. And then, what was weak, shallow, and unsatisfying, as it came from my mouth, shall and will be full, powerful, and satisfactory, as reinforced in thee by God's spirit.

DIALOGUE XII.

Means to be used by wounded Consciences for the recovering of Comfort.

Tim. Are there any useful means to be prescribed, whereby wounded consciences may recover comfort the sooner?

Phil. Yes, there are.

Tim. But now in the present day, some condemn all using of means: *let grace alone* (say they) fully and freely to do its own work, and thereby man's mind will in due time return to a good temper of its own accord: this is the most spiritual serving of God, whilst using of *means* makes but *dunces* and *truants* in *Christ's school*.

Phil. What they pretend spiritual will prove airy and empty, making lewd and lazy Christians: means may and must be used with these cautions. 1. That they be of God's appointment in his word, and not of man's mere invention. 2. That we still remember they are but means, and not the main. For to account of helps more than helps, is the highway to make them hindrances. Lastly, that none rely barely on the *deed done*, which conceit will undo him that did it, especially if any opinion of merit be fixed therein.

Tim. What is the first means I must use; for I reassume to personate a wounded conscience?

Phil. Constantly pray to God, that in his due time he would speak peace unto thee.

Tim. My prayers are better omitted than performed: they are so weak they will but bring the greater punishment upon me, and involve me within the *Prophet's* curse* to those *that do the work of the Lord negligently.*

Phil. Prayers negligently performed draw a curse, but not prayers weakly performed. The former is when one can do better, and will not; the latter is, when one would do better, but alas he cannot; and such failings as they are his sins, so they are his sorrows also: pray therefore faintly, that thou mayest pray fervently; pray weakly, that thou mayest pray strongly.

Tim. But in the law they were forbidden to offer to God any *lame* † sacrifice, and such are my prayers.

---

* Jer. xlviii. 10. † Deut. xv. 21.

Phil. 1. Observe a great difference betwixt the material *sacrifice* under the *law*, and spiritual Sacrifices (the *calves of the lips*) under the Gospel. The former were to be free from all blemish, because they did typify and resemble *Christ* himself. The latter (not figuratively representing *Christ*, but heartily presented unto him) must be as good as may be gotten, though many imperfections will cleave to our best performances, which by God's mercy are forgiven. 2. Know that, that in *Scripture* is accounted lame, which is counterfeit and dissembling, (in which sense hypocrites* are properly called halters) and therefore if thy prayer, though never so weak, be sound, and sincere, it is acceptable with God.

Tim. What other counsel do you prescribe me?

Phil. Be diligent in reading the word of God, wherein all comfort is contained; say not that thou art dumpish and indisposed to *read*, but remember how travellers must eat against their stomach: their journey will digest it; and though their palate find no pleasure for the present, their whole body will feel strength for the future. Thou

* 1 Kings xviii. 21.

hast a great journey to go, a *wounded conscience* has far to travel to find comfort, (and though weary, shall be welcome at his journey's end) and therefore must feed on God's word, even against his own dull disposition, and shall afterwards reap benefit thereby.

Tim. Proceed in your appointing of wholesome diet for my *wounded conscience* to observe.

Phil. Avoid *solitariness*, and associate thyself with *pious* and *godly company*. O the blessed fruits thereof! Such as want skill or boldness to begin or set a *Psalm*, may competently follow *tune* in concert with others. Many houses in *London* have such weak walls, and are so slightly and slenderly built, that were they set alone in the fields, probably they would not stand an hour; which, now ranged in *streets*, receive support in themselves, and mutually return it to others: so mayest thou in good society, not only be reserved from much mischief, but also be strengthened and confirmed in many godly exercises, which solely thou couldest not perform.

Tim. What else must I do?

Phil. Be industrious in thy calling. I press this

the more, because some erroneously conceive that a *wounded conscience* cancels all *Indentures of service,* and gives them (during their affliction) a dispensation to be idle. The inhabitants of the *Bishopric* of *Durham* * pleaded a privilege, that King *Edward the First* had no power, although on necessary occasion, to *press* them to go out of the *country*, because forsooth, they termed themselves *holy-work-folk*, only to be used in defending the holy *shrine* of St. *Cuthbert*. Let none in like manner pretend that (during the *agony* of a *wounded conscience*) they are to have no other employment than to sit moping to brood their melancholy, or else only to attend their devotions; whereas a good way to divert or assuage their pain within, is to take pains without in their vocation. I am confident that happy minute which shall put a period to thy misery shall not find thee idle, but employed, as ever some secret good is accruing to such who are diligent in their calling.

Tim. But though *wounded consciences* are not to be freed from all work, are they not to be favoured in their work?

* Cambd. Brit. in Durham.

Phil. Yes verily. Here let me be the advocate to such parents and masters who have sons, servants, or others, under their authority, afflicted with wounded consciences. O! do not with the Egyptian taskmasters, exact of them the full tale of their brick—*O spare a little till they have recovered some strength.* Unreasonable that maimed men should pass on equal duty with such soldiers as are sound.

Tim. How must I dispose myself on the Lord's day?

Phil. Avoid all servile work, and expend it only in such actions as tend to the sanctifying thereof. God, the great *Landlord* of all *time,* hath let out *six days* in the *week* to man to farm them; the *seventh day* he reserves as a *demesne* in his own hand: if therefore we would have quiet possession and comfortable use of what God hath *leased* out to us, let us not encroach on his *demesne.* Some *popish* people* make a superstitious *almanack* of the *Sunday,* by the fairness or foulness thereof guessing of the *weather* all the *week* after. But I

* If it rains on Sunday before Mess—it will rain all week more or less. A popish old rhyme.

dare boldly say, that from our well or ill spending of the *Lord's day*, a probable conjecture may be made how the following week will be employed. Yea, I conceive, we are bound (as matters now stand in *England*) to a stricter observation of the *Lord's day* than ever before. That a time was due to God's *service*, no Christian in our kingdom ever did deny: that the same was weekly dispersed in the *Lord's day, holy days, Wednesdays, Fridays, Saturdays*, some have earnestly maintained: seeing therefore all the last are generally neglected, the former must be more strictly observed; it being otherwise impious, that our devotion having a narrower *channel*, should also carry a shallower stream.

Tim. What other means must I use for expedition of comfort to my *wounded conscience?*

Phil. Confess* that sin or sins, which most perplexes thee, to some godly *minister*, who by absolution may pronounce, and apply pardon unto thee.

Tim. This confession is but a *device* of *Divines*, thereby to screw themselves into other men's secrets, so to mould and manage them with more ease to their own profit.

* 2 Sam. xii. 13; Matt. iii. 6.

Phil. God forbid they should have any other design but your safety; and therefore choose your confessor where you please, to your own contentment. So that you may find ease, fetch it where you may; it is not our credit, but your cure we stand upon.

Tim. But such confession hath been counted rather a rack for *sound* than a remedy for *wounded consciences.*

Phil. It proves so, as abused in the *Romish Church*, requiring an enumeration of all mortal sins, therein supposing an error, that some sins are not mortal, and imposing an impossibility, that all can be reckoned up. Thus the *conscience* is tortured, because it can never tread firmly, feeling no bottom, being still uncertain of confession, (and so of absolution) whether or no he hath acknowledged all his sins. But where this ordinance is commended as convenient, not commanded as necessary, left free, not forced in cases of extremity, sovereign use may be made, and hath been found thereof, neither *magistrate* nor *minister* carrying the *sword* or the *keys* in vain.

Tim. But Sir, I expected some rare inventions

G

from you for curing wounded consciences: whereas all your receipts hitherto are old, stale, usual, common, and ordinary; there is nothing new in any of them.

Phil. I answer, first, if a *wounded conscience* had been a *new disease*, never heard of in God's word before this time, then perchance we must have been forced to find out new remedies. But it is an *old malady*, and therefore *old physic* is best applied unto it. Secondly, the *receipts* indeed are old, because prescribed by him who is the *ancient of days*.* But the older the better, because warranted by experience to be effectual. God's ordinances are like the *clothes* † of the *children of Israel*: during our wandering in the wilderness of this world they never wax old, so as to have their virtue in operation abated or decayed. Thirdly, whereas you call them *common*, would to God they were so, and as generally practised as they are usually prescribed. Lastly, know we meddle not with *curious heads*, which are pleased with newfangled rarities, but with *wounded consciences*, who love solid comfort. Suppose our receipts ordinary

* Dan. vii. 9. † Deut. xxix. 5.

and obvious. If *Naaman** counts the cure too cheap and easy, none will pity him if still he be pained with his leprosy.

Tim. But your receipts are too loose and large, not fitted and appropriated to my malady alone. For all these (*pray, read, keep good company, be diligent in thy calling, observe the Sabbath, confess thy sins, &c.*) may as well be prescribed to one guilty of *presumption*, as to me ready to DESPAIR.

Phil. It doth not follow that our physic is not proper for one, because it may be profitable for both.

Tim. But despair and presumption being contrary diseases, flowing from contrary causes, must have contrary cures.

Phil. Though they flow immediately from contrary causes, yet originally from the common fountain of natural corruption; and therefore such means as I have propounded, tending towards the mortifying of our corrupt nature, may generally, though not equally, be useful to humble the presuming and comfort the despairing; but to cut

* 2 Kings v. 12.

off cavils in the next dialogue, we will come closely to peculiar counsels unto thee.

DIALOGUE XIII.

Four wholesome Counsels, for a wounded Conscience to practise.

Tim. Perform your promise: which is the first counsel you commend unto me?

Phil. Take heed of ever renouncing thy *filial interest* in *God*, though thy sins deserve that he should disclaim his *paternal relation* to thee. The *prodigal* * returning to his *father* did not say, *I am not thy son*, but, *I am no more worthy to be called thy son.* Beware of bastardizing thyself, being as much as Satan desires, and more than he hopes to obtain. Otherwise thy *folly* would give him more than his *fury* could get.

Tim. I conceive this a needful caution.

Phil. It will appear so if we consider what the Apostle † saith, that we *wrestle with principalities and powers.* Now *wrestlers* in the *Olympian*

* Luke xv. 21. † Ephes. vi. 12.

games were *naked,* and anointed with oil to make them sleek and glibbery, so to afford no holdfast to such as strove with them. Let us not gratify the Devil with this advantage against ourselves, at any time to disclaim our *son-ship* in God; if the Devil catches us at this lock, he will throw us flat, and hazard the breaking of our necks with final despair. Oh no! still keep this point; a *prodigal son* I am, but a *son,* no *bastard:* a *lost sheep,* but a *sheep,* no *goat :* an *unprofitable servant,* but *God's servant,* and not an absolute *slave to Satan.*

Tim. Proceed to your second counsel.

Phil. Give credit to what grave and godly persons conceive of thy condition, rather than what thy own fear (an incompetent judge) may suggest unto thee. A *seared conscience* thinks better of itself, a *wounded* worse, than it ought: the former may account all sin a sport, the latter all sport a sin. Melancholy men, when sick, are ready to conceit any cold to be the *cough* of the *lungs,* and an ordinary *Pustule* no less than the *plague sore.* So *wounded consciences* conceive sins of *infirmity* to be of *presumption,* sins of *ignorance* to be of *knowledge,* apprehending their case more dangerous than it is indeed.

Tim. But it seems unreasonable that I should rather trust another saying than my own sense of myself.

Phil. Every man is best judge of his *own self*, but during the swoon of a *wounded conscience* I deny thee to be come to *thy own self*: whilst thine eyes are blubbering, and a tear hangs before thy sight, thou canst not see things clearly and truly, because looking through a double *medium* of air and water; so whilst this cloud of *pensiveness* is pendent before the eyes of thy soul, thine estate is erroneously represented unto thee.

Tim. What is your third counsel?

Phil. In thy agony of a *troubled conscience* always look upwards unto a gracious God to keep thy soul steady, for looking downward on thyself thou shalt find nothing but what will increase thy fear, infinite sins, good deeds few, and imperfect: it is not thy faith, but God's faithfulness thou must rely upon; casting thine eyes downwards on thyself, to behold the great distance betwixt what thou deserveth and what thou desireth, is enough to make thee giddy, stagger, and reel into despair. Ever therefore *Lift up thine eyes unto the hills,* from whence cometh thy help,* never viewing the deep dale of thy own unworthiness but to abate thy pride when tempted to presumption.

Tim. Sir, your fourth and last counsel?

Phil. Be not disheartened as if comfort would not come at all, because it comes not all at once, but patiently attend God's leisure: they are not styled the swift, but the *sure* † *mercies of David,* and the same prophet says, *the glory of the Lord shall be thy rereward:* this we know comes up last to secure and make good all the rest: be assured, where *grace* patiently leads the *front, glory* at last will be in the *rear.* Remember the prodigious patience of *Elijah's* servant.

Tim. Wherein was it remarkable?

Phil. In obedience to his master: he went several times to the *sea;* it is tedious for me to tell what was not troublesome for him to do. 1, 2, 3, 4, 5, 6, 7 times sent down steep *Carmel,*‡ with

---

* Psalm cxxi. 1.   † Isaiah lv. 3, and lviii. 8.
‡ 1 Kings xviii. 43.

danger, and up it again with difficulty, and all to bring news of nothing, till his last journey, which made recompense for all the rest, with the tidings of a cloud arising. So thy thirsty soul, long parched with drought for want of comfort, though late, at last shall be plentifully refreshed with the *dew* of consolation.

Tim. I shall be happy if I find it so.

Phil. Consider the causes why a broken leg is incurable in a horse, and easily curable in a man: the horse is incapable of counsel to submit himself to the farrier, and therefore in case his leg be set, he flings, flounces, and flies out, unjointing it again by his misemployed mettle, counting all binding to be shackles and fetters unto him; whereas a man willingly resigns himself to be ordered by the *surgeon*, preferring rather to be a prisoner for some days, than a cripple all his life. *Be not like a horse or mule,* * which have no understanding ; but let patience † have its perfect work* in thee. When God goes about to *bind up* ‡ *the broken-hearted*, tarry his time, though ease come not at an instant; yea though it be painful for the

* Psalm xxxii. 9.　† James i. 3.　‡ Isa. lxi. 1.

present, in due time thou shalt certainly receive comfort.

DIALOGUE XIV.

Comfortable meditations for wounded consciences to muse upon.

Tim. Furnish me I pray with some comfortable *meditations*, whereon I may busy and employ my soul when alone.

Phil. First consider that our *Saviour* had not only a notional, but an experimental and meritorious knowledge of the pains of a *wounded conscience when hanging on the Cross.* If *Paul conceived himself happy* being to answer for himself, before King *Agrippa, especially because he knew* him to be expert in all the *customs and questions of the Jews*, how much more just cause has thy *wounded conscience* of comfort and joy, being in thy prayers to plead before *Christ* himself, who hath felt thy pain, and deserved that in due time by his stripes thou shouldest be healed?

Tim. Proceed I pray in this comfortable subject.

Phil. Secondly, consider that herein, like *Elijah*, thou needest not complain that thou art left *alone*, seeing the best of God's *saints* in all ages have smarted in the same kind—instance in *David* : indeed sometimes he boasts how *he lay in green* * *pastures, and was led by still waters;* but after he bemoans that *he sinks in deep* † *mire, where there was no standing.* What is become of those *green pastures?* parched up with the *drought.* Where are those *still waters?* troubled with the tempest of affliction. The same *David* compares himself to an *owl,*‡ and in the next *Psalm* resembles himself to an *eagle.* Do two *fowls* fly of more different kind? The one the *scorn,* the other the *sovereign;* the one the *slowest,* the other the *swiftest;* the one the most *sharp-sighted,* the other the most dim-eyed of all *birds.* Wonder not, then, to find in thyself sudden and strange alterations. It fared thus with all God's *servants*, in their agonies of temptation; and be confident thereof, though now run aground with grief, in due time thou shalt be all afloat with comfort.

* Psalm xxiii. 2.   † Psalm lxix. 2.
‡ Compare Psalm cii. 6, with Psalm cii. 5.

Tim. I am loth to interrupt you in so welcome a discourse.

Phil. Thirdly, consider that thou hast had, though not grace enough to cure thee, yet enough to keep thee, and conclude that he, whose goodness hath so long held thy head above water from drowning, will at last bring thy whole body safely to the shore. The *wife* of *Manoah* had more *faith* than her *husband*, and thus *she* reasoned : *If the Lord was pleased to kill us he would not have received a burnt and a meat offering at our hands.** Thou mayest argue in like manner : if God had intended finally to forsake me, he would never so often have heard and accepted my prayers, in such a measure as to vouchsafe unto me, though not full deliverance *from*, free preservation in my affliction. Know, *God hath done great things for thee already*, and thou mayest conclude, from his grace of supportation hitherto, grace of ease and relaxation hereafter.

Tim. It is pity to disturb you, proceed.

Phil. Fourthly, consider that, besides the private stock of thy own, thou tradest on the

* Judges xiii. 23.

public store of all good men's prayers, put up to heaven for thee. What a mixture of *Languages* met in *Jerusalem* at *Pentecost*,* *Parthians, Medes,* and *Elamites,* &c. But conceive to thy comfort, what a medley of prayers, in several tongues, daily centre themselves in God's ears in thy behalf, *English, Scotch, Irish, French, Dutch,* &c., insomuch, that perchance thou dost not understand one syllable of their prayers, by whom thou mayest reap benefit.

Tim. Is it not requisite, to entitle me to the profits of other men's prayers, that I particularly know their persons which pray for me?

Phil. Not at all, no more than it is needful that the *eye* or face must see the backward parts, which is difficult, or the inward parts of the body, which is impossible: without which sight, by sympathy they serve one another. And such is the correspondency by prayers, betwixt the mystical members of Christ's body, corporally unseen one by another.

Tim. Proceed to a fifth meditation.

Phil. Consider, there be five kind of *consciences*

* Acts ii.

on foot in the world: first, an *ignorant conscience*, which neither sees nor saith anything, neither beholds the sins in a soul, nor reproves them. Secondly, the *flattering conscience*, whose speech is worse than silence itself, which, though seeing sin, soothes men in the committing thereof. Thirdly, the *seared conscience*, which hath neither sight, speech, nor sense, in men that are *past feeling*.* Fourthly, a *wounded conscience*, frighted with sin. The last and best is a *quiet* and *clear conscience*, pacified in *Christ Jesus*. Of these, the fourth is thy case, incomparably better than the three former, so that a wise man would not take a world to change with them. Yea, a *wounded conscience* is rather painful than sinful, an affliction, no offence, and is in the ready way, at the next remove, to be turned into a *quiet conscience*.

Tim. I hearken unto you with attention and comfort.

Phil. Lastly, consider the good effects of a wounded conscience, privative for the present, and positive for the future. First, privative, this heaviness of thy heart (for the time being) is a

* Ephes. iv. 19.

bridle to thy soul, keeping it from many sins it would otherwise commit. Thou that now sittest sad in thy shop, or walkest pensive in thy parlour, or standest sighing in thy chamber, or liest sobbing on thy bed, mightest perchance at the same time be drunk, or wanton, or worse, if not restrained by this affliction. God saith in his prophet to Judah,* *I will hedge thy way with thorns*, namely, to keep Judah from committing *spiritual fornication*. It is confessed that a *wounded conscience*, for the time, is a *hedge* of *thorns* (as the messenger of *Satan*, sent to buffet St. *Paul*, is termed a *thorn* * *in the flesh*). But this *thorny* fence keeps our wild spirits in the true way, which otherwise would be straggling; and it is better to be held in the right road with *briars* and *brambles*, than to wander on beds of *roses*, in a wrong path, which leads to destruction.

Tim. What are the positive benefits of a wounded conscience?

Phil. Thereby the graces in thy soul will be proved, approved, improved. O, how clear will thy *sunshine* be when this *cloud* is blown over;

* Hos. ii. 6.   * 2 Cor. xii. 7.

and here I can hardly hold from envying thy happiness hereafter. O that I might have thy future *crown*, without thy present *cross*; thy *triumphs*, without thy *trial*; thy *conquest*, without thy combat! But I recall my wish, as impossible, seeing *what God hath joined together, no man can put asunder*. These things are so twisted together, I must have both or neither.

DIALOGUE XV.

That is not always the greatest sin whereof a man is guilty, wherewith his conscience is most pained for the present.

Tim. Is that the greatest sin in a man's soul, wherewith his *wounded conscience*, in the agony thereof, is most perplexed?

Phil. It is so commonly, but not constantly. Commonly indeed, that sin most pains and pinches him, which commands as principal in his soul.

Tim. Have all men's hearts some one *paramount* sin, which rules as *sovereign* over all the rest?

Phil. Most have. Yet as all countries are not *Monarchies,* governed by *Kings,* but some by *free states,* where many together have equal power; so it is possible (though rare) that one man may have two, three, or more sins, which jointly domineer in his heart, without any discernible superiority betwixt them.

Tim. Which are the sins that most generally wound and afflict a man when his *conscience* is terrified.

Phil. No general rule can exactly be given herein. Sometimes that sin, in acting whereof he took most delight; it being just, that the sweetness of his corporal pleasure should be sauced with more spiritual sadness. Sometimes that sin which (though not the foulest) is the most frequent in him. Thus his idle words may perplex him more than his oaths or perjury itself. Sometimes that sin, not which is most odious before God, but most scandalous before men, does most afflict him, because drawing greatest disgrace upon his person and profession. Sometimes that sin which he last committed, because all the circumstances thereof are still firm and fresh in his memory. Sometimes that sin, which (though long since by him committed) he hath heard very lately powerfully reproved; and no wonder if an old *gall*, new rubbed over, smart the most. Sometimes that sin, which formerly he most slighted and neglected, as so inconsiderably small, that it was unworthy of any sorrow for it, and yet now it may prove the sharpest sting in his conscience.

Tim. May one who is guilty of very great sins, sometimes have his conscience much troubled only for a small one?

Phil. Yes, verily: country patients often complain, not of the disease which is most dangerous, but most conspicuous. Yea, sometimes they are more troubled with the symptom of a disease (suppose an ill colour, bad breath, weak stomach) than with the disease itself. So in the soul, the conscience oft-times is most wounded, not with that offence which is, but appears most, and a sin incomparably small to others, whereof the party is guilty, may most molest for the present, and that for three reasons.

Tim. Reckon them in order.

Phil. First, that God may show in him, that as

sins are like the sands in number, so they are far above them in heaviness; whereof the least crumb taken asunder, and laid on the conscience, by God's hand, in full weight thereof, is enough to drive it to despair.

Tim. What is the second reason?

Phil. To manifest God's justice, that those should be choked with a *gnat-sin,* who have swallowed many *camel-sins,* without the least regret. Thus some may be terrified for not *fasting* on *Friday,* because indeed they have been *drunk* on *Sunday:* they may be perplexed for their wanton dreams when sleeping, because they were never truly humbled for their wicked deeds when waking. Yea, those who never feared *Babylon the Great,* may be frightened with *Little Zoar;* I mean, such have been faulty in flat superstition, may be tortured for committing or omitting a thing, in its own nature indifferent.

Tim. What is the third reason?

Phil. That this pain for a lesser sin may occasion his serious scrutiny into greater offences. Any paltry *cur* may serve to start and put up the *game* out of the bushes, whilst fiercer, and fleeter *hounds* are behind to course and catch it. God doth make use of a smaller sin, to raise and rouse the *conscience* out of security, and *to put it up*, as we say, to be *chased*, by the reserve of far greater offences, *lurking behind* in the soul, unseen, and unsorrowed for.

Tim. May not the conscience be troubled at that which in very deed is no sin at all, nor hath truly so much as but the appearance of evil in it?

Phil. It may. Through the error of the understanding such a mistake may follow in the conscience.

Tim. What is to be done in such a case?

Phil. The party's judgment *must be rectified*, before his conscience *can be pacified*. Then is it the wisest way to persuade him to lay the *axe* of repentance to the *root* of corruption in his heart. When real sins in his soul are felled by unfeigned sorrow, causeless scruples will fall of themselves. Till that root be cut down, not only the least *bough,* and *branch* of that *tree*, but the smallest *sprig, twig,* and *leaf* thereof, yea the very empty shadow of a *leaf* (mistaken for a sin, and created a fault by the jealousy of a misinformed judgment) is sufficient intolerably to torture a *wounded conscience*.

DIALOGUE XVI.

Obstructions hindering the speedy flowing of comfort into a troubled soul.

Tim. How comes it to pass, that comfort is so long in coming to some *wounded consciences?*

Phil. It proceeds from several causes, either from God, not yet pleased to give it; or the patient, not yet prepared to receive it; or the minister, not well fitted to deliver it.

Tim. How from God not yet pleased to give it?

Phil. His time to bestow consolation is not yet come: now no plummets of the heaviest human importunity can so weigh down God's *clock of time*, as to make it strike one minute *before his hour be come.* Till then, his *mother* herself could not prevail with *Christ*[*] to work a *miracle*, and turn *water into wine:* and till that minute appointed

[*] John ii. 4.

approach, God will not in a *wounded conscience* convert the *water of affliction* into that *wine of comfort*, which *makes glad the heart of the soul.*

Tim. How may the hindrance be in the patient himself?

Phil. He may as yet not be sufficiently humbled, or else God perchance in his providence foresees that as the *prodigal child*, when he had received his *portion*, riotously misspent it, so this *sick soul*, if comfort were imparted unto him, would prove an unthrift and ill husband upon it, would lose and lavish it. God therefore conceives it most for his glory, and the other's good, to keep the comfort still in his own hand, till the *wounded conscience* get more wisdom to manage and employ it.

Tim. May not the *sick man's* too mean opinion of the minister be a cause why he reaps no more comfort by his counsel?

Phil. It may. Perchance the sick man hath formerly slighted and neglected that minister, and God will now not make him the instrument for his comfort, who before had been the object of his contempt. But on the other side, we must also know, that perchance the party's over-high opinion of the minister's parts piety, and corporal presence (as if he cured where he came, and carried ease with him), may hinder the operation of his advice. For God grows jealous of so suspicious an instrument, who probably may be mistaken for the principal. Whereas a meaner man, of whose spirituality the patient hath not so high carnal conceits, may prove more effectual in comforting, because not within the compass of suspicion to eclipse God of his glory.

Tim. How may the obstructions be in the minister himself?

Phil. If he comes unprepared by prayer, or possessed with pride, or unskilful in what he undertakes; wherefore in such cases, a minister may do well to reflect on himself (as the *disciples** did when they could not cast out the devil), and to call his heart to account, what may be the cause thereof: particularly whether some unrepented-for sin in himself hath not hindered the effects of his counsel in others.

Tim. However, you would not have him wholly disheartened with his ill success?

* Matt. xvii. 19.

Phil. Oh no; but let him comfort himself with these considerations. First, that though the patient gets no benefit by him, he may gain experience by the patient, thereby being enabled more effectually to proceed with some other in the same disease. Secondly, though the sick man refuses comfort for the present, yet what doth not sink on a sudden, may soak in by degrees, and may prove profitable afterwards. Thirdly, his unsucceeding pains may notwithstanding facilitate comfort for another to work in the same body, as *Solomon* built a temple with most materials formerly provided, and brought thither by *David.* Lastly, grant his pains altogether lost on the *wounded conscience,* yet his *labour is not in vain in the Lord,** who, without respect to the event, will reward his endeavours.

Tim. But what if this minister hath been the means to cast this sick man down, and now cannot comfort him again?

Phil. In such a case, he must make this sad accident the more matter for his humiliation, but not for his dejection. Besides, he is bound, both

* 1 Cor. xv. 58.

in honour and honesty, civility and christianity, to procure what he cannot perform, calling in the advice of others more able to assist him, not conceiving, out of pride or envy, that the discreet craving of the help of others is a disgraceful confessing of his own weakness: like those malicious *midwives*, who had rather that the woman in travail should miscarry, than be safely delivered by the hand of another, more skilful than themselves.

DIALOGUE XVII.

What is to be conceived of their final estate who die in a wounded Conscience *without any visible Comfort.*

Tim. What think you of such, who yield up their ghost in the agony of an *afflicted spirit*, without receiving the least sensible degree of comfort?

Phil. Let me be your remembrancer to call to or keep in your mind, what I said before, that our discourse only concerns the children of God: this notion renewed, I answer. It is possible that the sick soul may receive secret solace, though the standers-by do not perceive it. We know how insensibly Satan may spirt and inject despair into a heart, and shall we not allow the Lord of heaven to be more dexterous and active with his antidotes than the devil is with his poisons?

Tim. Surely, if he had any such comfort, he would show it by words, signs, or some way, were it only but to comfort his sad kindred, and content such sorrowful friends who survive him; were there any hidden fire of consolation kindled in his heart, it would sparkle in his looks and gestures, especially seeing no obligation of secrecy is imposed on him, as on the blind man,* when healed, to tell none thereof.

Phil. It may be he cannot discover the comfort he hath received, and that for two reasons. First, because it comes so late, when he lies in the *meshes* of life and death, being so weak, that he can neither speak nor make signs with *Zechariah;* being at that very instant *when the silver cord is ready to be loosed, and the golden bowl to be broken, and the pitcher to be broken at the fountain, and the wheel to be broken at the cistern.*

* Mark viii. 26.

Tim. What may be the other reason?

Phil. Because the comfort itself may be incommunicable in its own nature, which the party can take, and not tell; enjoy, and not express; receive, and not impart: as by the assistance of God's spirit, he sent up *groans which cannot be uttered,** so the same may from God be returned with comfort which cannot be uttered; and as he had many invisible and privy pangs, concealed from the cognizance of others, so may God give him secret comfort, known unto himself alone, without any other men's sharing in the notice thereof. *The heart knoweth his own bitterness; and a stranger doth not intermeddle with his joy.*† So that his comfort may be compared to the *new name* given to God's servants, *which no man knoweth, save he that receiveth it.*‡

Tim. All this proceeds on what is possible or probable, but amounts to no certainty.

Phil. Well, then, suppose the worst: this is most sure, though he die without tasting of any comfort here, he may instantly partake of everlasting joys hereafter. Surely many a despairing soul, groan-

---
* Rom. viii. 26.   † Prov. xiv. 10.   ‡ Rev. ii. 17.

ing out his last breath with fear and thought to sink down to hell, hath presently been countermanded by God's goodness to eternal happiness.

Tim. What you say herein, no man alive can confirm or confute, as being known to God alone, and the soul of the party. Only I must confess, that you have charity on your side.

Phil. I have more than charity, namely, God's plain and positive promise, *blessed are such as mourn, for they shall be comforted.** Now, though the particular time, *when,* be not expressed, yet the latest *date* that can be allowed, must be in the world to come, where such mourners, who have not felt God in his comfort here, shall see him in his glory in Heaven.

Tim. But some who have led pious and godly lives have departed, pronouncing the sentence of condemnation upon themselves, having one foot already in hell by their own confession.

Phil. Such confessions are of no validity, wherein their fear *bears false witness* against their faith. The fineness of the whole cloth of their life must not be thought the worse for a little coarse *list* at

* Matt. v. 4.

the last. And also their final estate is not to be construed by what was dark, doubtful, and desperate at their deaths, but must be expounded by what was plain, clear, and comfortable, in their lives.

Tim. You, then, are confident that a holy life must have a happy death?

Phil. Most confident. The *Logicians* hold that, although from false premises a true conclusion may sometimes follow, yet from true propositions nothing but a truth* can be thence inferred; so though sometimes a bad life may be attended with a good death (namely, by reason of repentance, though slow, sincere, though late, yet unfeigned, being seasonably interposed), but where a godly and gracious life hath gone before, there a good death must of necessity follow; which, though sometimes doleful (for want of apparent comfort) to their surviving friends, can never be dangerous to the party deceased. Remember what *St. Paul* saith, *Our life is hid with Christ in God.*†

Tim. What makes that place to your purpose?

* *Ex veris possunt, nil nisi vera sequi.*   † Col. iii. 3.

Phil. Exceeding much. Five cordial observations are couched therein. First, that God sets a high price and valuation on the souls of his servants, in that he is pleased to *hide* them : none will *hide* toys and trifles but what is counted a treasure. Secondly, the word *hide*, as a relative, imports that some seek after our souls, being none other than *Satan* himself, *that roaring lion, who goes about* SEEKING, *whom he may devour.** But the best is, let him *seek*, and *seek*, and *seek*, till his malice be weary (if that be possible) : we cannot be hurt by him whilst we are *hid in God.* Thirdly, grant *Satan* find us there, he cannot fetch us thence : *our souls are bound in the bundle of life, with the Lord our God.* So that, be it spoken with reverence, God first must be stormed with force or fraud, before the soul of a saint sinner, *hid* in him, can be surprised. Fourthly, we see the reason why so many are at a loss, in the agony of a *wounded conscience,* concerning their spiritual estate. For they look for their life in a wrong place, namely, to find it in their own piety, purity, and inherent righteousness. But though they

* 1 Peter v. 8.

seek, and search, and dig, and dive never so deep, all in vain. For though *Adam's* life was hid in himself, and he intrusted with the keeping his own integrity, yet, since Christ's coming, all the original evidences of our salvation are kept in a higher office, namely, *hidden in God himself.* Lastly, as our *English proverb* saith, *he that hath hid can find;* so God (to whom *belongs the* issues from death*) can infallibly find out that soul that is hidden in him, though it may seem, when dying, even to labour to lose itself in a fit of despair.

Tim. It is pity, but that so comfortable a doctrine should be true.

Phil. It is most true: surely as *Joseph* † and *Mary* conceived that they had lost *Christ* in a crowd, and sought him three days sorrowing, till at last they found him beyond their expectation, safe and sound, sitting in the *Temple:* so many pensive parents, solicitous for the souls of their children, have even given them for gone, and lamented them lost (because dying without visible comfort) and yet, in due time, shall find them to

---
* Psalm lxviii. 20. † Luke ii. 48.

their joy and comfort, safely possessed of honour and happiness, in the *midst of the heavenly temple,* and *church triumphant* in glory.

DIALOGUE XVIII.

Of the different time and manner of the coming of comfort to such who are healed of a wounded conscience.

Tim. How long may a servant of God lie under the burden of a *wounded conscience?*

Phil. It is not for us to know the times and the seasons which the Father hath put in his own power.* God alone knows whether their grief shall be measured unto them, by hours, or days, or weeks, or months, or many years.

Tim. How, then, is it that *St. Paul* saith, that *God will give us the issue with the temptation,*† if one may long be visited with this malady?

Phil. The apostle is not so to be understood, as if the *temptation* and *issue* were *twins,* both borne at the same instant; for then no affliction could

* Acts i. 7. † 1 Cor. x. 13.

last long, but must be ended as soon as it is begun; whereas we read how *Aneas** truly *pious* was bedridden of the *palsy* eight years: the *woman* diseased with a *bloody issue* † twelve years; another *woman bowed by infirmity* ‡ eighteen years; and the man § *lame* thirty-eight years at the *pool of Bethesda.*

Tim. What, then, is the meaning of the Apostle?

Phil. God *will give the issue with the temptation;* that is, the *temptation* and the *issue* bear both the same date in God's decreeing them, though not in his applying them: at the same time, wherein he resolved his servants shall be tempted, he also concluded of the means and manner, how the same persons should infallibly be delivered. Or thus: *God will give the issue with the temptation;* that is, as certainly, though not as suddenly. Though they go not *abreast,* yet they are joined successively, like two links in a chain, where one ends the other begins. Besides there is a twofold *issue;* one, through a *temptation;* another out of a *temptation.* The former is but mediate, not

---

* Acts ix. 33. † Matt. ix. 20. ‡ Luke xiii. 11.
§ John v. 5.

final; an *issue* to an *issue*, only supporting the person tempted for the present, and preserving him for a future full deliverance. Understand the *apostle* thus, and the *issue* is always both given and applied to God's children, with the *temptation*, though the temptation may last long after, before fully removed.

Tim. I perceive then, that in some a wounded conscience may continue many years.

Phil. So it may. I read of a poor *widow*, in the land of *Limburgh**, who had nine children, and for thirteen years together was miserably afflicted in mind, only because she had attended the dressing and feeding of her little ones before going to *mass*. At last it pleased God to sanctify the endeavours of *Franciscus Junius*, that learned, godly *divine*, that upon true information of her judgment she was presently and perfectly comforted.

Tim. Doth God give ease to all in such manner on a sudden?

Phil. O no: some suddenly receive comfort, and in an instant they pass from *midnight* to bright

* Melchior Adamus invita Theologorum Extororum, page 198.

day, without any dawning betwixt. Others receive consolation by degrees, which is not poured, but dropped into them by little and little.

Tim. Strange, that God's dealing herein should be so different with his servants.

Phil. It is to show that as in his proceedings there is *no variableness,** such as may import him mutable or impotent, so in the same there is very much variety, to prove the fulness of his power and freedom of his pleasure.

Tim. Why doth not God give them consolation all at once?

Phil. The more to employ their prayers and exercise their patience. One may admire why *Boaz*† did not give to *Ruth* a quantity of corn more or less, so sending her home to her mother, but that rather he kept her still to glean; but this was the reason, because that is the best charity which so relieves another's poverty as still continues their industry. God, in like manner, will not give some consolation all at once; he will not spoil their (painful but) pious profession of gleaning; still they must pray and gather, and

* James i. 17. † Ruth ii. 8.

pray and glean, here an ear, there a handful of comfort, which God scatters in favour unto them.

Tim. What must the party do when he perceives God and his comfort beginning to draw nigh unto him?

Phil. As *Martha*,* when she heard that *Christ* was coming, stayed not a minute at home, but went out of her house to meet him, so must a sick soul when consolation is coming, haste out of himself, and hie to entertain God with his thankfulness. The best way to make an *Omer* of comfort increase to an *Ephah* (which is ten times as much †) is to be heartily grateful for what one hath already, that his store may be multiplied. He shall never want more who is thankful for and thrifty with a little: whereas ingratitude doth not only stop the flowing of more mercy, but even spills what was formerly received.

* John xi. 20.   † Exod. xvi. 36.

DIALOGUE XIX.

How such who are completely cured of a wounded conscience are to demean themselves.

Tim. Give me leave now to take upon me the person of one recovered out of a wounded conscience.

Phil. In the first place, I must heartily congratulate thy happy condition, and must rejoice at thy *upsitting*, whom God hath raised from the bed of despair: welcome *David* out of the *deep*, *Daniel* out of the *lions' den*, *Jonah* from the *whale's belly*: welcome *Job* from the *dunghill*, restored to health and wealth again.

Tim. Yea, but when *Job's brethren* came to visit him after his recovery, every one gave him a *piece of money and an ear-ring of gold* :* but the *present* I expect from you, let it be, I pray, some of your good counsel for my future deportment.

Phil. I have need to come to thee, and comest thou to me? Fain would I be a *Paul* sitting at the feet of such a *Gamaliel*, who hath been cured of a

* Job xlii. 11.

wounded conscience in the height thereof. I would turn my *tongue* into *ears*, and listen attentively to what tidings he brings from hell itself. Yea, I should be worse than the brethren of *Dives*, if I should not believe one risen from the *dead*, for such in effect I conceive to be his condition.

Tim. But, waiving these digressions, I pray proceed to give me good advice.

Phil. First, thankfully own God thy principal restorer and comforter paramount. Remember that of *ten Lepers* * one only returned to give thanks, which shows that by nature, without grace over-swaying us, it is ten to one if we be thankful. Omit not also thy thankfulness to good men, not only to such who have been the architects of thy comfort, but even to those who though they have built nothing, have borne burthens towards thy recovery.

Tim. Go on, I pray, in your good counsel.

Phil. Associate thyself with men of afflicted minds, with whom thou mayest expend thy time to thine and their best advantage. O how excellently did *Paul* comply with *Aquila* and *Priscilla*!

* Luke xvii. 17

As their hearts agreed in the general profession of piety, so their hands met in the trade of *tent makers*;* they abode and wrought together, being of the same occupation. Thus I count all *wounded consciences* of the same company, and may mutually reap comfort one by another: only here is the difference: they (poor souls) are still bound to their hard task and trade, whilst thou (happy man) hast thy *indentures cancelled*, and being *free* of that *profession*, art able to instruct others therein.

Tim. What instructions must I commend unto them?

Phil. Even the same comfort, wherewith thou thyself was comforted of God:† with *David*, tell them *what God hath done for thy soul*; and with *Peter*, being strong, *strengthen thy brethren*:‡ conceive thyself, like *Joseph*, therefore, *sent before*, and *sold* into the *Egypt* of a *wounded conscience* (*where thy feet were hurt in the stocks, the irons entered into thy soul*), that thou mightest provide *food* for the *famine* of others, and especially be a purveyor of comfort for those thy brethren, which

* Acts xviii. 3. † 2 Cor. i. 4. ‡ Luke xxii. 32.

afterwards shall follow thee down into the same doleful condition.

Tim. What else must I do for my afflicted brethren?

Phil. Pray heartily to God in their behalf: when *David* had prayed, *Psalm* xxv. 2, *O my God, I trust in thee, let me not be ashamed;* in the next verse (as if conscious, to himself, that his prayers were too restrictive, narrow, and niggardly) he enlarges the bounds thereof, and builds them on a broader bottom; *yea, let none that wait on thee be ashamed.* Let charity in thy devotions have *Rechoboth,* room enough: beware of pent petitions, confined to thy private good, but extend them to all God's servants, but especially all *wounded consciences.*

Tim. Must I not also pray for those servants of God which hitherto have not been *wounded* in *conscience?*

Phil. Yes, verily, that God would keep them from or cure them in the exquisite torment thereof. Beggars, when they crave an alms, constantly use one main motive, that the person of whom they beg may be preserved from that misery, whereof they themselves have had woeful experience. If they be blind, they cry, *master, God bless your eyesight;* if lame, *God bless your limbs;* if undone by casual burning, *God bless you and yours from fire.* Christ, though his person be now glorified in heaven, yet he is still subject by sympathy of his saints on earth, to hunger, nakedness, imprisonment, and a *wounded conscience*, and so may stand in need of feeding, clothing, visiting, comforting, and curing. Now, when thou prayest to Christ for any favour, it is a good plea to urge, edge, and enforce thy request withal, *Lord, grant me such or such a grace, and never mayest thou, Lord, in thy mystical members, be tortured and tormented with the agony of a* wounded conscience, *in the deepest distress thereof.*

Tim. How must I behave myself for the time to come?

Phil. Walk humbly before God, and carefully avoid the smallest sin, always remembering Christ's caution;* *behold thou art made whole: sin no more, lest a worse thing come unto thee.*

* John v. 14.

DIALOGUE XX.

Whether one Cured of a wounded conscience be subject to a Relapse.

Tim. May a man, once perfectly healed of a *wounded conscience*, and for some years in peaceable possession of comfort, afterwards fall back into his former disease?

Phil. Nothing appears in Scripture or reason to the contrary, though examples of real relapses are very rare, because God's servants are careful to avoid sin, the cause thereof, and being once burnt therewith, ever after dread the fire of a *wounded conscience*.

Tim. Why call you it a relapse?

Phil. To distinguish it from those *relapses* more usual and obvious, whereby such who have snatched comfort before God gave it them, on serious consideration that they had usurped that to which they had no right, fall back again into the former pit of despair; this is improperly termed a relapse, as not being a renewing, but a continuing of their former malady, from which, though seemingly, they were never soundly recovered.

Tim. Is there any intimation in Scripture of the possibility of such a real relapse in God's servants?

Phil. There is; when *David* saith, Psalm lxxxv. 8, *I will hear what God the Lord will speak, for he will speak peace unto his people, and to his Saints, but let them not turn again to folly:* this imports that if his *Saints turn again to folly,* which by woeful experience we find too frequently done, God may *change his voice,* and turn his peace, formerly spoken, into a warlike defiance to their conscience.

Tim. But this, methinks, is a diminution to the majesty of God, that a man, once completely cured of a wounded conscience, should again be pained therewith. Let *mountebanks* palliate, cures break out again, being never soundly, but superficially healed; *he that is all in all* never doth his work by *halves,* so that it shall be undone afterward.

Phil. It is not the same individual wound in number, but the same in kind, and, perchance, a deeper in degree: nor is it any ignorance, or falsehood in the surgeon, but folly and fury in the patient, who by committing fresh sins, causes a new pain in the old place.

Tim. In such relapses men are only troubled for such sins which they have run on score since their last recovery from a wounded conscience.

Phil. Not those alone, but all the sins which they have committed both before and since their conversion may be started up afresh in their minds and memories, and grieve and perplex them with the guiltiness thereof.

Tim. But those sins were formerly fully forgiven, and the pardon thereof solemnly sealed and assured unto them; and can the guilt of the same recoil again upon their consciences?

Phil. I will not dispute what God may do in the strictness of his justice. Such seals, though still standing firm and fast in themselves, may notwithstanding break off and fly open in the feeling of the sick soul: he will be ready to conceive with himself that as *Shimei,** though once forgiven his railing on *David*, was afterwards executed for the same offence, though upon his

* 1 Kings ii. 44.

committing of a new transgression, following his servants to *Gath*, against the positive command of the king, so God, upon his committing of new trespasses, may justly take occasion to punish all former offences; yea, in his apprehension, the very foundation of his faith may be shaken, all his former title to heaven brought into question, and he tormented with the consideration that he was never a true child of God.

Tim. What remedies do you commend to such souls in relapses?

Phil. Even the self-same receipts which I first prescribed to *wounded consciences*, the very same promises, precepts, comforts, counsels, cautions. Only as *Jacob*, the second time that his *sons* went down into *Egypt*,* commanded them to carry double money in their hands, so I would advise such to apply the former remedies with double diligence, double watchfulness, double industry, because the malignity of a disease is riveted firmer and deeper in a relapse.

* Genesis xliii. 12.

DIALOGUE XXI.

Whether it be lawful to pray for, or to pray against, or to praise God for a wounded conscience.

Tim. Is it lawful for a man to pray to God to visit him with a *wounded conscience?*

Phil. He may and must pray to have his high and hard heart truly humbled and bruised with the sight and sense of his sins, and with unfeigned sorrow for the same: but may not explicitly and directly pray for a *wounded conscience*, in the highest degree and extremity thereof.

Tim. Why interpose you those terms *explicitly* and *directly?*

Phil. Because *implicitly*, and by consequence, one may pray for a *wounded conscience:* namely, when he submits himself to be disposed by God's pleasure, referring the particulars thereof wholly to his infinite wisdom, tendering, as I may say, a blank paper to God in his prayers, and requesting him to write therein what particulars he pleases; therein generally, and by consequence, he may pray for a *wounded conscience*, in case God sees the same for his own glory and the party's good; otherwise, directly he may not pray for it.

Tim. How prove you the same?

Phil. First, because a *wounded conscience* is a judgment, and one of the sorest, as the resemblance of the torments of hell. Now, it is not congruous to nature or grace for a man to be a free and active instrument, purposely to pull down upon himself the greatest evil that can befall him in this world. Secondly, we have neither direction, nor precedent of any *saint*, recorded in God's word, to justify and warrant such prayers. Lastly, though praying for a *wounded conscience* may seemingly scent of pretended humility, it doth really and rankly savour of pride, limiting *the holy one of Israel;* it ill becoming the patient to prescribe to his heavenly physician what kind of physic he shall minister unto him.

Tim. But we may pray for all means to increase grace in us, and therefore may pray for a *wounded conscience*, seeing thereby, at last, piety is improved in God's servants.

Phil. We may pray for and make use of all

means whereby grace is increased: namely, such means as by God are appointed for that purpose; and therefore, by virtue of God's institution, have both a proportionableness and a tendency in order thereunto. But, properly, those things are not means, or ordained by God, for the increase of piety, which are only accidentally overruled to that end by God's power, against the intention and inclination of the things themselves. Such is a *wounded conscience*, being always actually an evil of punishment, and too often occasionally an evil of sin, the *bias* whereof doth bend and bow to wickedness; though, overruled by the aim of God's eye and *strength of his arm*, it may bring men to the mark of more grace and goodness. God can and will extract *light* out of *darkness*, *good* out of *evil*, *order* out of *confusion*, and *comfort* out of a *wounded conscience:* and yet *darkness, evil, confusion*, &c., are not to be prayed for.

Tim. But a *wounded conscience*, in God's children, infallibly ends in comfort here or glory hereafter, and therefore is to be desired.

Phil. Though the ultimate end of a *wounded conscience* winds off in comfort, yet it brings with it 'many intermediate mischiefs and maladies, especially as managed by human corruption: namely, dulness in divine service, impatience, taking God's name in vain, despair for the time, blasphemy; which a saint should decline, not desire; shun, not seek; not pursue, but avoid with his utmost endeavours.

Tim. Is it lawful positively to pray against a *wounded conscience?*

Phil. It is, as appears from an argument taken from the lesser to the greater. If a man may pray against pinching poverty, as wise *Agur* did,* then may he much more against a *wounded conscience*, as a far heavier judgment. Secondly, if God's servants may pray for ease under their burthens, whereof we see divers particulars in that worthy prayer of *Solomon*,† I say, if we pray to God to remove a lesser judgment by way of subvention, questionless we may beseech him to deliver us from the great evil of a *wounded conscience*, by way of prevention.

Tim. May one lawfully praise God for visiting him with a *wounded conscience?*

* Proverbs xxx. 8.   † 1 Kings viii. 33.

Phil. Yes, verily. First, because it is agreeable to the will of God* *in everything to be thankful:* here is a general rule, without limitation. Secondly, because the end why God makes any work is his own glory; and a *wounded conscience* being a work of God, he must be glorified in it. especially seeing God shows much mercy therein, as being a punishment on this side of *hell fire*, and less than our deserts. As also, because he hath gracious intentions towards the sick soul for the present, and when the malady is over the patient shall freely confess that it is *good for him that he was so afflicted.* Happy, then, that soul, who in the *lucid intervals* of a wounded conscience can praise God for the same. *Music* is sweetest near, or over rivers, where the *echo* thereof is best rebounded by the water. Praise for pensiveness, thanks for tears, and blessing God over the floods of affliction makes the most melodious music in the ear of heaven.

* 1 Thes. v. 18; Ephes. v. 20; Psalm ciii. 22; and cxlv. 10.

THE CONCLUSION OF THE AUTHOR
TO THE READER.

AND now God knows how soon it may be said unto me, *physician, heal thyself,* and how quickly I shall stand in need of these counsels which I have prescribed to others. Herein I say, with *Eli* to *Samuel*,* *it is the Lord, let him do what seemeth him good:* with *David* to *Zadock*,† *behold here I am, let him do to me as seemeth good unto him.* With the *Disciples* to *Paul*,‡ *the will of the Lord be done.* But oh, how easy it is for the mouth to pronounce or the hand to subscribe these words! But how hard, yea, without God's grace, how impossible, for the heart to submit thereunto! Only hereof I am confident, that the making of this *treatise* shall no ways cause or hasten a *wounded conscience* in me, but rather on the contrary (especially if, as

* 1 Sam. iii. 18. † 2 Sam. xv. 26. ‡ Acts xxi. 14.

it is written *by* me, it were written *in* me) either prevent it, that it come not at all, or defer it, that it come not so soon, or lighten it, that it fall not so heavy, or shorten it, that it last not so long. And if God shall be pleased hereafter to write *bitter things against me,** who have here written the sweetest comforts I could for others, let none insult on my sorrows. But whilst my *wounded conscience* shall lie, like the *cripple,*† at the *porch of the temple,* may such as pass by be pleased to pity me, and permit this book to beg in my behalf the charitable prayers of well-disposed people; till divine providence shall send some *Peter,* some pious *minister,* perfectly to restore my maimed soul to her former soundness. *Amen.*

* Job xiii. 26.   † Acts iii. 2.

FINIS.

TRIANA;

OR, A THREEFOLD

ROMANZA,

Of { MARIANA,
     PADUANA,
     SABINA.

Omne tulit Punctum qui miscuit utile dulci.

WRITTEN

BY THO. FULLER, B.D.

LONDON:
PRINTED FOR JOHN STAFFORD, AND ARE TO BE SOLD
AT HIS HOUSE AT THE GEORGE, AT
FLEETBRIDGE. 1664.

TO THE READER.

It is hard to say whether it is worst to be Idle, or ill employed; whilst I have eschewed the former, I have fallen on the latter, and shall by the severer sort, be censured for misspending my time.

Let such, I pray hear my *Plea*, and I dare make my accusers my Judges herein; that is not lost time, which aims at a good end. *Sauce is as lawful as meat, recreation as labour;* it hath pleased me in composing it, I hope it may delight others in perusing it.

I present not a translation out of the *Spanish*, or from the *Italian Original;* this is the common *Pander* to men's fancy, hoping to vent them under that title, with the more applause. These my play-labours never appeared before, and is an essay of what hereafter may be a greater volume.

Things herein are composed in a general proportion to truth, and we may justly affirm, *Vera si non scribimus, scribimus veri similia.*

I will not be deposed for the exact variety of the gravest passages. In the greatest historian a *liberty* hath ever been allowed to fancies of this nature, always provided, that they confine themselves within the bounds of probability.

Thus wishing every faithful lover, *Feliciano's* happiness; every good wife, *Facundo's* love; every true servant, *Gervatio's* fortune; every maiden lady, *Fidelio's* constant affection; every faithful friend, *Vejeto's* success; every clownish fool, *Insuls'* mishap; and every cruel wanton, *Nicholayo's* deserved punishment, I leave thee to the perusal hereof. Censure not so rigidly, lest you blast a budding writer, in the blossoming of his endeavours.

TRIANA.

MARIANA.

IN the City of *Valentia, Metropolis* of the *Kingdom* so named, which, with many other Dominions, are the tributary Brooks discharging themselves into the ocean of the *Spanish Monarchy*, dwelt one *Don Durio*, a merchant of great repute. For as yet, the envious sands had not, (as at this day) obstructed the Haven in *Valentia*, but that it was the principal port in those parts.

This *Don Durio* had advanced an estate, much by Parsimony, more by Rapine, being half a Jew by his *extraction*, and more than three quarters thereof by his *Conditions,* being a notorious oppressor. But grown very aged, and carrying his *Eyes* in his *Pocket, Teeth* in his *sheath,* and *Feet* in his *Hands,* he began with remorse to reflect on the former part of his Life, with some thoughts of restitution to such whom he had most injured.

This his intention, he communicated to one *Francisco*, a Friar, and his *Confessor*. *Francisco* was glad to see such a qualm of Religion come over his heart, and resolved to improve it to the uttermost. He persuades him that restitution was a thing difficult and almost impossible for one in his condition, so many were the particular persons by him wronged. The shortest and surest way, was for him to consign his only Daughter *Mariana*, to be a Nun in the Priory of *St. Bridget*, and to endow that Convent with all his Lands; which exemplary piece of his liberality, would not only with the lustre thereof, outshine all his former faults, but also be a direction to posterity, how to regulate their estates on the like occasion. *Don Durio*, though flinty of himself, yet lately softened with age and sickness, entertains the motion, not only with contentment, but with delight, and will not be a day older, before the same be effected.

But there was a material person, whose consent herein must be consulted with, even *Mariana* his

Daughter, who had not one ounce of *Nun's* flesh about her; as whom nature had intended, not as a dead stake in a hedge, to stand singly in the place, but as a plant, to fructify for posterity. Besides, she had assured herself to one *Fidelio*, a Gentleman of such merit, that though his virtues started with great disadvantage, clogged with the weight of a necessitous fortune; yet such the strength and swiftness thereof, that he very speedily came (not being above the years of two and twenty,) to the mark of a public reputation. But these things were carried so closely between them, and all leaks of superstition were so cunningly made up, that neither friend nor foe, had gained the least glimpse of their intentions.

Don Durio, Francisco being in his presence, importunes his Daughter, (a hard task,) to bury her beauty under a veil, and become a *Bridgettine.*

What he propounded with a fatherly bluntness, *Francisco* sharpens with the edge of his wit, heightening the happiness of a secluded Life to the Sky, and above it; a discourse very unwelcome to *Mariana's* ears, racketted between two dangers on either side. If she surrender herself herein to her Father's will, she is undone; and what she values above herself, *Fidelio* is ruined. If she deny, she exposeth herself to the just censure of disobedience: yea it puts a light into the hand of her suspicious Father, thereby to discover her intentions, that her affections being pre-engaged, obstructed the acceptance of this motion. No time is allowed her to advise. In a moment, (almost,) she must consult and conclude, and resolved at last to comply with her Father's desires for the present, not despairing but that courteous time, in the process thereof, would tender unto her some advantage, whereby hereafter, she might make a fair evasion.

But her Father hurries her in her present attire, (as good enough for a mortified mind, without allowing her respite of exchanging,) unto the Convent. *Francisco* leads the way, *Don Durio* follows, and *Mariana* comes last. Her countenance was neither so sad, as to betray any discontent, nor so blithe and cheerful, as to proclaim any likeness therein; but so reduced, and moderately composed, as of one that well understood both what she was leaving off, and what she was entering into. And if the falling of a few tears moistened her cheeks, it was excusable in one now taking her farewell of her former friends; and her Father beheld the same, as the Argument of good nature in her.

Ringing the bell at the *Convent* Gate, the watchful Porter takes the Alarm, and presently opens; for though it was something difficult for strangers to have access into the Convent, yet the presence of Friar *Francisco*, was as strong as any *Petard*, to make the sturdiest gate in the *Convent* pliable to his admission. Out comes the Lady *Abbess*, who had now passed sixty winters, and carried the repute of a grave and sanctimonious Matron. A strict discipliner she was, of the least wantonness of any committed to her charge; reputed by most to her virtuous disposition, but ascribed by others to her envy, driving away those delights from others, which formerly had flown away from herself.

Francisco with a short speech, acquaints her with the cause of their coming, surrenders *Mariana* to be a Probationer in their house, whom the *Abbess* welcometh with the largest expressions of love to her and thankfulness to her Father; highly commending *Don Durio's* Devotion, that whereas many Parents, blessed with a numerous issue, grudge to bestow the tithe thereof on a Monastical Life, such is his forward zeal, as to part with all his stock and store, not repining to confer his sole daughter and heiress, to a religious retirement. Then taking their leaves each of other, they depart, leaving *Mariana* with the *Abbess*.

Millicent, a Nun of good esteem and great credit, is assigned by the Abbess to go along with *Mariana*, and show her all the rooms, walks, and rich utensils of the house; especially she is very careful to read unto her a large Inventory of all the relics therein, with their several miraculous operations. The points of *St. Rumball's* Breeches, (among other things,) were there shown, the touching whereof, would make barren women fruitful; and many other seeming toys of sovereign influence. Passing through by the south-east corner of the Cloister, *Mariana* cast her eye on an arched vault, enquiring the use thereof, and whither the same did conduct.

Millicent answered, that in due time her curiosity therein should be satisfied to her own contentment; but as yet she was not capable of any intelligence therein, which was one of the mysteries of the house, not communicated to novices at their first admission, but reserved for such, who after some convenient time of abode there, had given undoubted testimony of their fidelity to that Order.

And here we leave *Mariana*, having more music and less mirth, than she had at home.

The news hereof was no sooner brought to *Fidelio*, but it moved a strange impression upon him. Were I assured that the reader hereof, was ever found in love, and that his breast was ever through warmed with chaste fires of a constant mistress, it would spare me some pains to character *Fidelio's* sad condition. For then my work is easily done, only, by appealing to the reader's experience; who out of a sympathy, is able to make more than a conjecture of *Fidelio's* sad estate, daily languishing for the loss of his Love as dead, (whilst living,) unto him.

His pensive postures, sad looks, silent sighs,

affected solitariness, sequestering himself from his most familiar friends, was observed by *Ardalio*, by whom he was entirely beloved. *Ardalio* using the boldness becoming a friend, examines *Fidelio* of the cause of his sudden change. *Fidelio* for a time, fences himself with his own retiredness, and fortifies his soul, with resolutions of secrecy. The other, plants the Artillery of his importunity against him, by the force whereof, *Fidelio* is beaten out of his hold, and won at last, to unbosom his grievance to *Ardalio*; who had promised, that *Midnight* sooner should be found a *Tell-tale*, and *Trust* itself become a Traitor, rather than he would discover anything prejudicial unto him. Emboldened wherewith, *Fidelio* confesseth, that *Mariana's* restraint in a Nunnery, into which she was lately thrust by her parent's power, (as he verily believed,) against her own will, and without his knowledge.

Had she been taken prisoner by the *Turk*, some hope would have been to procure her *Liberty* by ransom; had *Pirates* surprised her, money might have purchased her freedom; whereas now no hope of enlargement, it being no less than Sacrilege accounted, by force, or fraud, to practise her delivery from that Religious slavery.

Be content, (saith *Ardalio*,) and I will make you master of a project, which without any danger, shall bring your Mistress into your possession. Turn, therefore, all your Soul into ears, and listen to my discourse, which though seemingly tedious in the relating, the same will make your attention, not only a favour, but a gainer by the rich conclusion thereof.

Some twelve years since, when a Friar was buried in the Benedictine Convent, all the School-boys in Valentia, (among whom I assure you I was none of the meanest,) invited themselves to be present at the solemnity. Now, whilst others were crowded into the Chapel, to see the performance of the Obsequies, I know not what conceit made me a separatist from the rest of my companions. And as I was walking, in a Corner between the Conduit and the Hall, I happened into an old Room, which led me into a Vault, lighted only with one squint-eyed Window: going somewhat further therein, my heart began to fail me, with the fresh remembrance of those

Tales of Bugbears, wherewith my Nurse had affrighted my infancy from ponds and places of dangers: however, taking heart, I resolved to discover the issue of that winding Vault. And here, you must forgive me, if I have forgotten some circumstances herein. My memory, which never was very loyal, may be pardoned for betraying some passages, after twelve years past. Let it suffice, that I remember so much, as will make you happy if wisely prosecuted. This was the result of my adventure,—that, as our River *Anas* is reported to run some miles under ground, and afterwards spring up again, so I, drowned under dry earth, (if you will allow the expression,) was boiled up again in the Nunnery of the Bridgetines. You will hardly believe with what amazement the Nuns beheld me, who had entered their Cloister that unusual way, never as yet, (as it seems,) traced with boys' feet, but by those of more maturity, whose company might be more acceptable unto them. They loaded me with kisses and sweetmeats, for, believe me (*Fidelio*), how mean opinion so ever you may now have of my handsomeness, if my picture then taken, and the report of my Mother may be believed, I was not unhandsome. In fine, they flattered and threatened me, not to discover which way I came thither; which if once I made known, thousands of Devils would torment. From which day to this, it never came into my head: my memory having now made some amends for its former forgetfulness by this seasonable suggestion thereof unto me, when it may befriend your occasions. If therefore, you can convey yourself by this passage into the Nunnery, I leave the rest to be stewarded by your own Ingenuity.

But which way (said *Fidelio*) shall I contrive my undiscovered coming thither? Show me but a way presenting but half the face, yea, but a quarter face of probability, and I shall kiss that, and thank you for the same.

I will furnish you with all requisites for the Adventure. I have an uncle living at *Lisbon*, provincial of the Benedictines in Spain: his hand, when a boy, I have often counterfeited (for harmless cheats level with my age), so livelily, that I have persuaded him to confess it his own; and great the familiarity between my Uncle and

Francisco, who here is the Benedictine Abbot. I will provide all things for you, and fit you with the habit of their Order, leaving the rest still to be improved, by your own Art and Industry. For where a friend tenders one hand to draw you out of the Mire, if you assist not to make up the rest by your own Endeavours, even lie there still, to your own shame, and with no pity of mine.

They depart, resolving next morrow to meet, when all necessaries should be provided. In the mean time *Fidelio* goes to the Convent of the Bridgetines, and walks under a Window thereof, the which (as he was informed, and his intelligence therein not untrue), belonged to the Chamber of *Mariana*. She discovers him there, and presently rends out of her legend the first white leaf which had not blushed (as the rest), for the lies and impudences which were written therein; which *Paper* she employs for a *Letter*, and looking out of the window, casts the letter down unto him who stood ready beneath, to receive the same; and surely, had the *Letter* been but balanced with any competent weight put therein, it had not missed his hands for whom it was intended.

What a pity was it, that *Æolus* was never in love, or that the *Winds* are too boisterous or too cold to be melted by *Affection*, except any will say that a Gale of wind was ambitious to kiss the letter of so fair a hand, and overacted its part therein.

Sure it is, that a small blast thereof, blew this Letter over into the Garden of the *Abbess*, where she was viewing of her Bees. Her Ladyship betakes herself to her glass eyes, and peruseth the following contents thereof.

FIDELIO,

Help me with thy imaginations, and know me here more miserable than I can express; here is nothing less than that which is pretended, a chaste mistress which in due time may be a chaste wife, may stock a hundred Nuns with Virginity; work my deliverance, if thy affections be unfeigned, or I am undone.

MARIANA.

Short and sweet, said the *Abbess*, the least *Toads* have the greatest *Poison*. And then up she flies, winged with anger, (which otherwise could scarce crawl,) to *Mariana's* Chamber, where she so rails on her, that a Purgatory hereafter, might have been spared for having one here.

And because she had defamed the whole Convent, the heaviest penance must be enjoined unto her, to be stripped naked to the middle, in the *Hall* at dinner time, where she was only to be feasted with lashes, each *Nun* inflicting one upon her, and then, the *Abbess* to conclude *sans* number, as many as her own discretion was pleased to lay on her.

But *Mariana*, partly with grief, and partly with fear, fell so terribly sick, that that night her life came into despair; the only reason (as I take it), why her penance was put off to another time, till she might be the more able object of their cruelty. For the *Abbess* resolved, that what was deferred should not be taken away; being so far from abating the principal, she intended *Mariana* should pay it with interest, and give satisfaction for the forbearance of this Discipline, when in any tolerable strength to undergo it.

By this time *Ardalio* had completed *Fidelio*

with all necessaries in the habit of a Friar, who thus accoutred, advanceth in his formality to the Convent, where he is presently brought to *Francisco*, to whom he delivered this ensuing Letter.

DEAR BROTHER,

I send you here myself in my friend, who, was he as well known to you as to me, his virtues would command your affection, if not admiration. Our Convent hath this last hundred years, (since the first foundation therof,) been essaying and endeavouring to make up a complete man, which now in some measure it had effected in the bearer hereof; when behold, envious death, repining at our happiness, had laboured to frustrate the same; so that this peerless piece of devotion was scarce bestowed on us, when almost taken away from us by a violent sickness, whose abated fury terminated at last in a long and languishing quartan ague, which his hollow eyes and thin cheeks do too plainly express. Exchange of Air is commended unto us, for the best, if not only Physic. Let him want nothing, I pray, your house can afford, and burthen him with as few questions as may be, it being tedious to him to talk, and his infirmity hath much disabled his intellectuals; and know that your meritorious kindness therein shall not only oblige me to a requital, but put an engagement on all Lovers of virtue, to whom this our Brother *Festuca's* worth is or hereafter may be known.

Your loving Friend,
Pedro di Ronca.

Francisco embraceth him with all dearness; the viands of the house are set before him, whereof *Fidelio* took a moderate repast; all the Monks of the house severally salute him, and demanded of him several particulars of their Convent at *Lisbon*, as of the Situation, Endowment, Number, Names, and punctual observances therein.

To all these *Fidelio* returned general answers, under the cover whereof he might the easier conceal his own ignorance: as for the endowment of the house, he professed himself a mere stranger thereunto, and that he affected ignorance therein, as conceiving it resented too much of worldly employment, whereas his desire was, that better things should engross his soul. This excuse was heard with admiration, increasing the opinion they had preconceived of his holiness. When they ask him such questions to which he could return no answer at all, he would fall into a seeming trance, darting his eyes, and moving his lips as in pious ejaculations, as not listening to what was said unto him; whilst the others, out of mannerly devotion, let fall their questions, and would press them no farther, as loth to interrupt his soul in more serious and sacred employment. Besides, *Francisco* gave a strict command, that none should disturb him with needless questions, but leave him to the full employment of his own meditations.

Some six days after, *Fidelio*, observing the directions of *Ardalio* his friend, finds out the aforesaid vault, and accordingly went forward therein, till at last he came to a Great Iron Gate, which was shut, and obstructed the way. This Gate was either not set up, or not shut up, when *Ardalio* made his passage this way. *Fidelio* falls a musing, finds all his hopes prove abortive, with no possibility of further proceeding, when looking seriously on the Gate, which was enlightened through a small hole, presenting but twilight at noonday, he discovered an inscription thereon, which he read to himself, with a voice one degree above whispering.

> Fair Portress of this privy Gate,
> If any Sister want a Mate,
> Lift up the Bar and let me in,
> It shall be but a venial sin.

Instantly the Gate flew open, and *Fidelio* is left to wonder what secret Spell and hidden Magic were contained in those verses, that so immediately they should procure his admittance. But presently the Riddle is unfolded, finding a *Nun* behind the door, where she kept her constant attendance, and to whom these verses were the wonted watch-word to open the Gate. Re-collecting his spirit, he accosted the Sister, desiring to be conducted to *Mariana*, whom he understood (for the fame thereof had gotten out of the Convent), was very sick, and he, sent from Father *Francisco*, to provide some Ghostly counsel for her, and what else her present condition should require.

He is carried unto her Chamber, who being at the present asleep, he entertained himself in the next room by looking on a *Picture*. Herein, Saint *Denis* was set forth, beheaded by *Pagan Tyrants*, and afterwards carrying his head under his arm, seven miles from *Paris* to Saint *Denis*. *Fidelio*, smiling thereat to himself, thought, that Saint *Denis* in that posture, had an excellent opportunity to pick his teeth.

Mariana is awaked, *Fidelio* called in, who, claiming the privilege of privacy, as proper for a Confessor, all avoid the room; then putting off his disguise, he briefly telleth her what dangers he had undergone for her deliverance, informeth her that next morning he would be ready with horses and servants, at such a corner of the Garden, where, without any danger, they had contrived a way for her escape. *Fidelio's* presence is better than any Cordial. *Mariana* recovereth her spirits, is apprehensive of the motion, promiseth to observe time and place, they are thrifty in their language, speak much in little, lavish no minutes in compliments, but are perfectly instructed in each other's designs.

Presently, in comes the *Abbess*. *Fidelio* (having first recovered his cowl), proceeds in a set discourse, of the praise of patience, commending it to *Mariana* as the most necessary virtue in her condition; and after some general salutes to the *Abbess*, recommendeth his *Patient* to her care, and takes his own opportunity to depart, returning to the Convent of the *Benedictines* as undiscovered as he came thence.

That night, befriended with the dark, under the mantle thereof, *Fidelio* gets out of the Convent, repairs to his friend *Ardalio*, reports all the particulars of his success, triumphs in the hopes of his approaching happiness, counts the time (which we generally complain doth fly), a Cripple that crawls, so long each minute seems unto him, until ten of the clock next morning should be accomplished.

Next day Mariana riseth from her bed, craves leave of the *Abbess* to go into the Garden; which is granted her, so that *Millicent*, a prime *Nun*, were to attend her. *Millicent* adviseth her to wrap herself warm for fear of a relapse, for the Air (saith she) is piercing, and your body weak.

Indeed, (quoth *Mariana*), I am sensible of much strength in a short time, and believe myself able to run a race with you, who shall first come to yonder corner. Done, said *Millicent*, and let the lag that comes last to the place, say over her Beads for us both. *They start together.*

Have you ever heard the Poet's fiction of *Atalanta's* running, when only her covetousness to get the golden Ball made her lose the race? Alas! her swiftness was nothing in comparison of *Mariana's*. It seems that *Cupid*, who had shot the piles of his arrows into her heart, had tied the feathers or wings thereof unto her feet.

The wall of the *Nunnery* was on the one side filled up with Earth; on the outside there was a descent of some four yards. *Fidelio* stood ready to entertain her. *What will not fright and love do?* *Mariana* consults not any danger, nor did she carry a scale in her eye to measure the depth of the wall; but crossing the Proverb, *she leaps before she looks*, *Fidelio* saveth her almost half the way of her journey, by catching of her in his arms, and before the succeeding minute had supplanted its Predecessor, she is mounted by his servant behind *Fidelio*, and all speed they make to a Chapel, which might serve for an *Hospital*, for it was blind, and the Priest dumb, yet had tongue enough to tie that knot which none of them was able to undo.

But tired *Millicent*, with much panting, had recovered the corner, and seeing *Mariana's* escape, cried out, *she is gone, she is gone*. This gave an *Alarm to the Abbess*, who instantly conceived her in a swoon. It is, said she, the just reward of her adventure, that contrary to my Counsel went abroad. Out comes the whole Regiment of *Nuns*, with hot waters and Cordials, to tempt her soul, if not too far departed, to return to her body. They are soon sensible of their mistake, and behold her almost got out of sight, so that the longer they looked the less they saw her, till distance at last made her vanish away. Many a *Nun* a *Spectator* hereof, wished themselves an *Actor* upon the same terms, commending her adventure in their hearts, who condemned her most in their discourse.

Fidelio, with *Mariana* his wife, returned to a private house in *Lisbon*, prepared for this purpose.

But oh! the quick intelligence that *Friars* have! Sure their souls are all scent, all Eyes and Ears. that discover things so soon, so far off. They were just ready to sit down to supper, only expecting the coming of his dear friend *Ardalio*, when in comes a man, or a Tiger (shall I say). Nature might seem to intend him for the latter, such his fierce aspect and hairy face, the terribleness whereof, was increased with his affected antic attire. By his place, he was the *Jailor of the house of Inquisition*, and presently he showed a cast of his office, by seizing them both his Prisoners. Yet, might they have had the happiness to have been sent to the same Prison, it would have afforded some mitigation of their misery. This would not be granted, though *Mariana* with importunate tears requested it; but they were disposed into several Jails, where neither of them was sensible of their own condition, being totally taken up with the mutual bemoaning the one, of the other.

This is one commendable quality in the *Spaniards;* prisoners are not long delayed to rot in the Jail, where is life worse than death itself, but are brought to a speedy trial, either to be condemned or acquitted. Next day, they both are brought before the *Judges*, and condemned to die; he, for Sacrilege, for soliciting a votary out of the house; she, for carrying away a golden Medal, wherein was the Picture of Saint *Bridget*, which she casually borrowed, having no felonious intention, as meaning to restore it, but surprised on a sudden, had no leisure to make restitution. All conceived that the rigours was extended unto them, and by *Francisco* the *Friar*, though not visibly appearing.

Don Durio hearing that his Daughter was to be executed, his Paternal affections retreated to his heart, and there made a *Stand*, projecting with himself how to prevent this mischief. And here I must trouble the Reader to go back in reporting an accident that happened twenty years ago.

It chanced that *Philip* the second, (always wealthy and always wanting,) was forced on a sudden, to send forth a great Fleet against the *Turk*. He borrowed a considerable sum of money of *Don Durio*, for the payment whereof, *Don Durio* was a daily suppliant to the Court, as constant at the gate as the Porter, plying the King with importunate Petitions, all which ended in delays, which *Don Durio* rightly expounded to be denials. Once the Treasurer told him that it was honour enough for the greatest *Monarch* in *Christendom* to borrow money of him, though he never receive it again. *Don Durio*, to make a virtue of necessity, turned his despair into a frolic; and being admitted by friends into the King's presence on a Winter's morning, cast into the fire his obligations, which were parcelled up in a pretty bundle, desiring the King to heat his hands thereat. *His Majesty* was highly pleased with the Conceit, and the rather, because it was more than a conceit; saying it was the best Faggot he ever saw, and wished the State of *Senua* would make him the like Bonfire; swearing by Saint *James*, (his usual Oath), that if ever *Don Durio* had need of a Court favour at a dead lift, he should not fail in his expectation.

The dead lift, (or at least the dying lift), was now come. *Don Durio* posts to *Madrid*, where the *Spanish Court* was kept, and findeth the *King* hunting of a stag The old man attends the sport for a time. The stag wearied with long hunting took foil, and ran into a great Pond or *dwarf Lake*. He recruited his strength, as old *Æson* did in the Bath of *Medea*, and came forth as fair and as fresh, as when roused in the morning; then setting his Haunches against the Park pale, (Reader, if a *Forester*, pardon my language if improper), he dared the *Dogs* to set upon him. Whilst the hounds stood disputing with themselves, (for the King's dogs we know can make syllogisms), whether the honour, or the danger were the greater to adventure their Foe, and whilst they stood declining the hazard one to the other, out steps a cowardly keeper, and with a brace of bullets killeth the Stag dead in the place ; who, could he have borrowed a tongue from the standers by, first he would have cursed that *Friar* of *Mentz*, for first finding out the hellish invention of Gunpowder, and then, he would have bequeathed himself to have been coffined in paste, whilst the *King* and his Courtiers should be merry at the solemnizing of his Funeral.

The sport being ended, *the King* returned and retired to his chamber. *Don Durio* makes his address to his *Majesty*, who at the first had forgotten him, till his memory was quickened with the effectual token of the Bonds he burnt. Welcome *Woodmonger*, said the *King*, thy suit is granted in the asking of it; and presently a large pardon was signed and sealed, which with all possible speed he carried along with him to *Lisbon*.

But so short the day, so long the way, so bad the weather, that he could not make such speed as he desired and his friends expected.

The day of execution being come, *Fidelio* is brought to act his part on death's *Theatre*. *Mariana*, though disjoined from him in Prison, to her great grief, was now, to her greater grief, conjoined with him at the Scaffold. *Fidelio* begins with a long speech, which seems no whit tedious to the Auditors, because done out of a design to gain time, in expectation of a pardon, which all understood was procured. All Lovers there present could have wished each vowel long in his speech, the effect whereof was to advise young persons to confine their affections within some probable compass of their deserts, not to wander with their extravagant love above the proportion of their merits. He bemoaneth himself much, *Mariana* more, taking on himself the guilt of the whole Action, and protested that she died Love's true Martyr.

Mariana seconded him in this sad discourse, the purport whereof was to teach obedience to Children, that they should take heed how they concealed their Love from their Parents, in whose mere disposal they were; and not to conceive that Age superannuated them, or gave them an acquittance from that debt to which nature engaged them.

A *Post* winds his horn; all hear it and welcome, conceiving what indeed it was, the preface to a pardon. *Don Durio* follows the Post all in a sweat, it being almost a wonder that his dried corpse could contribute so much moisture. The pardon is presented to the *Supreme Officer*, with much joy and acclamation of the beholders.

Who would think that *Heraclitus* could be so soon turned into *Democritus?* Who could suppose that so great an Army of people could in an instant Face about? It was hard before to find one merry, now impossible to find one sad; as if by sympathy they had been condemned with *Fidelio*, and accordingly pardoned with him.

The pardon is read: it was large parchment, in character, but apprehended too narrow in expression, as only for the life of *Mariana*, whose Father, *Don Durio*, neither desired nor endeavoured the life of the other, whom he perfectly hated, as conceiving his love a disparagement.

Writers were in a sad condition if sometimes they might not take upon trust from their readers more than they are able to pay themselves; how short would he fall, who would undertake in language, to express the general sadness of the Company, but especially of *Mariana*, for this unexpected accident. The Executioner proceeds to his work, a handkerchief being tied about *Fidelio's* face, as one better prepared to feel, than see death; he is readily provided for the fatal stroke.

In vain did *Mariana* with much Rhetoric (grief making her eloquent), plead that the pardons of *Princes* are not to be taken in restrictive senses; that in all things which are doubtful, men are rather to enlarge it with favour, than contract it with cruelty; that though her pardon alone was expressed, doubtless both were intended; that man and wife were but one, the guilt but one, committed by both; and appeals to the judges present, if any spark of mercy were alive in their breasts (Judges always for the greater solemnity being present at Executions), to improve the same on so just, so conscientious, so honourable an occasion. But as soon might a Child have persuaded the Tide at full sea to retreat, when enraged also with the wind, as her request find any entertainment.

Ardalio was present thereat, standing close to Father *Francisco*, the great Actor herein, who spurred on the *Judges* (whom charity otherwise believed inclined to mercy, to the greatest and speediest extremity), and he desired a private word with *Francisco*. What was whispered between them was unknown, and men's Fancies variously commented on their discourse, but the truth was, he spake to this effect.

Sir, you have been the grand *Engineer* of this man's death, whose blood you have sought, being yourself guilty of greater offences. A word from your mouth may respite the execution, and reprieve the Prisoner. I protest revenge of my friend's blood, if you do not quickly improve your utmost: three minutes is all the time I allow you to think, or do, after I have ended my speech. Know you, Sir, a Vault and a Door, between your Convent and the Nuns', contrary to Canons and Laws Ecclesiastical and Civil. These things shall be heightened against you, with as much earnestness as my wit and wealth can improve it, intending to bury my estate in the prosecution of the death of my friend. These things he uttered with that seriousness which protests no passion, but a calm soul, and such people are truest to their resolutions.

Francisco went to the prime Officer, and requested him (Friars' requests in such cases being commands), to put off the execution for one month, until his Majesty's pleasure therein was more perfectly known; for *Ardalio* had given him private information, that the intention of the King was larger than his expression in the pardon; and the officer complied with him in his desire, and all the Company were dissolved, none being sad at so strange but unexpected an alteration.

All matters were hushed and stopped; *Ardalio* embraced and feasted by Francisco, who bribed his tongue to silence, which the other as ingenuously professed and faithfully performed. The reprieve of *Fidelio* ended in a full pardon, and old *Don Durio*, seeing it to be in vain to forbid that match which providence had made, was contented that his Daughter was enjointred in a true affection, and consented unto their Marriage. Both which lived long, and were blessed with a happy posterity.

FINIS.

TRIANA AND PADUANA

In the City of Venice there flourished a Merchant as large in estate, as narrow in heart, (*Mellito Bondi* by name), of a Family more ancient than numerous, and yet more numerous than rich, until *Mellito* gave the lustre thereunto with the vastness of his estate. One daughter alone he had, *Paduana*, (from the neighbouring place of her birth and breeding), courted by all the *Illustrissimos* and *Clarissimos* of that State, as well she might, having the portion of a Princess in expectation. Yet her wealth was the meanest thing about her, whose virtues and beauty were such, that fame, commonly a Liar in the excess, was here a liar in the defect, her large report falling short of the Lady's due deserts.

Paduana, solicited to marriage, denied all suitors, charging all upon the account of her steadfast resolutions on virginity; whereas, this was but a *blind* more covertly to conceal her affections, and that exchange of hearts, which had passed between her and *Feliciano*, a sojourner in the house of her Father.

This *Feliciano* was a proper Gentleman, completely educated, whose enemies allowed this to be his worst fault, that he had a prodigal to his Father, who had wasted the large estate of his Ancestors; yet let not old *Andrea* (for so was his Father termed) be wholly condemned for an unthrift, (the partial cause of his ruin), seeing that losses at Sea, and ill debtors at land, contributed to his speedier undoing; besides, our aforesaid *Bondi*, if strictly examined, could not deny his concurrence thereunto, who by usurious contracts and sinister advantages, spurred him on to destruction, who was running too fast of himself. Hereupon, in some commiseration, he kept *Feliciano*, his son, as a gentle Almsman, exhibiting diet and some slender accommodations unto him.

The best was, the scant measure of *Bondi's*

allowance was enlarged by his daughter's bounty, maintaining him in a fashionable *equipage*. Thus for a time we leave them to their embraces, so much the sweeter because the secreter, waiting the leisure of every opportunity, and warily stealing the same.

It happened about this time, that the *President* of *Dalmatia* languished on a desperate sickness, his death being daily expected. This was an office of great honour and expense, which could not be creditably discharged without the annual expending of so many ducats, which amount to three thousand pounds sterling of our English money; for though the aforesaid *President* had a pension from the State, and a certain *Intrado* from the Galleys and Garrisons, with some considerable revenues from the demesnes annexed to the place, yet all his perquisites and emoluments audited, the aforesaid sum was requisite to carry it forth with any reputation, except some sordid soul was careless of his credit, and would sacrifice the same to public obloquy.

Mellito Bondi was designed by the *Duke and Senate of Venice* successor in this Presidentship of

Dalmatia. Indeed, in Seniority it belonged unto him; and as it was accounted an injury to balk so good ground, and pass over a man of merit, should the *State* decline his election; so, on the other side, it would have left an indelible shame to *Mellito* if he should have waived the acceptance thereof.

Mellito quakes for fear to be advanced downward to so chargeable a preferment; his covetousness is above his ambition, and he almost dies for fear, to hear that the President of *Dalmatia* is dying. The news of the arriving of a wealthy Ship from *Cairo*, or *Constantinople*, would be far more acceptable unto him than such burthensome honour.

Now, he had a confidant, part Friend, part servant, *Gervatio* by name, whose secrecy he had bought, and whose tongue he had locked up with many favours bestowed upon him. To him he presumes to impart his grievance in manner following.

Gervatio, I rank my servants in a threefold Order, of Slaves, Servants, and Friends: of the former I have many fit for servile labour, no ingenious employment; of the middle sort I want none, but these love mine rather than me; of friends-servants thou art the chief. I make thy own ingenuity my judge, whether my carriage unto thee hath not rather spoke me a Father to a Child, rather than a Master to a Servant. Thankful natures (among whom I shall account you, till discerning the contrary), will study to deserve favours bestowed upon them, which begetteth in me a confidence that I may not only safely trust thee with an important secret, but also, crave thy advice therein for my further direction.

Gervatio made a short, but serious protest of his fidelity herein, professing himself highly advanced in this trust committed unto him, withal much commending *Bondi's* ability to advise himself; bemoaning withal his own insufficiency, who could not harbour so presumptuous a thought, as if the scant measure of his weak judgment could supply anything wanting in the rich Treasury of his Master's experience. However, he promised that his heart should make recompense for his head, and the sincerity of his endeavours make some amends for his other failings and infirmities.

From compliments, I fall to the matter in hand. *Bondi* tells him his great desire to decline the costly *Presidentship of Dalmatia*, which by succession, when vacant, was certain to descend upon him; he voweth that he accounted it ill husbandry to sell rich lands, therewith to buy empty air and honourable titles, which vanish with the wearer thereof, whilst his lasting wealth might probably descend to his posterity, and desireth *Gervatio* to mind him of some fair contrivance, which might not leave the blur of any suspicion behind it (much less come within the reach of probable detection), whereby he might waive that expensive place, though but for a time, until some other should be settled in the same.

Presently *Gervatio's* better *genius* prompted him, that any person how rich soever, if totally deprived of any of his senses, was by the fundamental Laws of that State rendered incapable of the aforesaid *Presidentship*. On this undoubted founda ion, being a most certain and well-known truth, he bottomed his design, persuading *Bondi* to counterfeit himself stark blind; that this infirmity, cunningly dissembled and generally believed, would secure him out of the distance of the danger he feared, being not eligible to the place while visited with so great a defect.

The plot takes with *Bondi*, who puts it in present execution. On the morrow an entertainment is made, some friends invited to celebrate the anniversary of his birthday, and *Bondi* proves himself a perfect Miser by his over-plentiful cheer. In the midst of their mirth he complains the Room is dark; commands the windows to be opened, which was done accordingly. *Bondi* perseveres in his complaint, that he sees no more than he did before, (which in some sense was not untrue). At last, all means used to recover light unto him prove in vain, so that *Justice* is not fancied more blind by the *Philosophers*, nor *Cupid* feigned more eyeless by the Poets, than *Bondi* was then believed to be.

This accident produced different effects, as men stood variously affected unto him. Narrow the number of such who truly loved him, and those few did really pity and bemoan him; more his foes, who rejoiced thereat, and made uncharitable constructions thereof, as if some secret wanton intemperance had deprived him of his sight: none suspected any fraud or collusion therein. And to make all the surer, *Bondi* confessed that this was a just punishment inflicted upon him for his pride and ambition, because he so greedily had desired the *Presidentship of Dalmatia*. He acknowledged that he had been no better than a murderer in his own heart, having often killed the old *President of Dalmatia* in his wishes and desire, the sooner to pave the way to his own preferment, and ennoble his posterity with the addition of so honourable employment.

Hitherto *Feliciano* and *Paduana* had managed their affections with all secrecy, suffering none to be privy thereunto; but henceforward, being ignorant of her Father's dissimulation, they abstain not in his presence to pass kisses and courtesies, as confident that he perceived nothing; whereat the other was enraged above measure. Should his daughter, being a fit match for a Prince, for Parts, Portion, and Pedigree, be cast away on the son of a Bankrupt, all whose maintenance proceeded out of his own purse? He resolved rather to disinherit *Paduana* than endure this affront, though for the present in silence he digested the same.

The long-languishing *President of Dalmatia* puts an end of people's expectations by his death, and made room for one to succeed him in his office. The election leaves over *Bondi*, by his blindness unqualified for that place; the concernment whereof, required one who should be an *Argus* both for body and mind, such is the need of his constant wariness and circumspection. *Martino Carnatio* is by general suffrages reputed for the place, legally chosen, and solemnly settled therein, and conducted to *Spolato* by the Galleys of the state, where he began his residence, and we meet with no further mention of him.

Soon after the *Duke of Venice* comes to give *Bondi* a visit, bringing along with him a Chirurgeon, whose skill as it was diffused over all sores, so his masterpiece was in being an expert *oculist*. Indeed the eye is a small volume, but many the leaves (I mean the tunicles), thereof, and much written therein, the eye alone being subject to more distempers and diseases than any other part of the Body, so many and so small the contrivances therein, and no wonder if little Watches be quickly out of order.

This *Oculist* (being indeed one of more fame than skill, of more skill than honesty), at the *Duke's* desire, made an accurate inspection of *Bondi's* eyes, and pretended that he discovered in both of them a little speck hindering the sight thereof, which with a small Instrument might easily be cut out, with very little pain; and here he scattered a multitude of hard and long Latin words, which would serve for the titles of the Gallipots of an Apothecary's shop, which much amused his hearers therewith. I spare the pains to relate them, because questioning the reader's skill in understanding thereof, the rather because I question the *Oculist*, whether he understood himself in them.

Bondi was now put to it, fearing some violence should be offered to his eyes, where a touch is a wound, such the tenderness thereof. He thanks the *Chirurgeon* much, and the *Duke* more, for their care and kindness unto him; but is resolved patiently to bear the affliction laid upon him, which he confesses himself did justly deserve for his pride and ambition: he hopes his soul should be the better for the blindness of his body, and in no case would consent that any experiment should be tried on him for his recovery.

Here the *Duke* interposed his power and flat command. *Signor Bondi*, said he, you are not so in your own full and free dispose, but you may and must be overruled by others for your own good. We take notice of that worth in you which your modesty will not own in yourself, and therefore will not lose the benefit of so useful a patriot. You are a self murderer, if refusing those lawful means whereby Art may befriend Nature against your infirmity. As your friend, therefore, I desire you, as your Prince I command you, as both I enjoin you, without further dispute to submit yourself to this Artist, not doubting but that his learned endeavours will be crowned with welcome success.

Wit works in extremity. Now or never *Bondi* play thy prizes. With a composed countenance, he returns.

My Lord, I am ready with all thankfulness to embrace your counsel, and much admire the extensiveness of your goodness and largeness of your spirit, that amidst the multiplicity of your state employments, your ability is such, and your goodness so great, as you will reflect on so inconsiderable a thing as myself. But give me leave to acquaint your grace, I have lately made a vow to my particular Saint, whose aid I have implored (and whose name for some reasons I crave leave as yet to conceal), and have received some assurance from him in a dream, that shortly I shall be bettered by his goodness. I call it a dream, but surely it was not such, wherein Fancy commands in the absence of reason; but certainly, my Lord, such was the impression thereof in my soul, that it carrieth with it the presage of somewhat more than ordinary. Favour me then only to expect the issue thereof, and if my dream be but a dream, I shall yield myself wholly up to be ordered by your grace's pleasure, and thankfully accept what course soever shall be prescribed unto me. Hereof the Duke was contented, and after the exchange at some compliments, the company departed.

Next day *Bondi* calls *Feliciano*, and sends him to the shrine of Saint *Sylvester* in *Ancona*, desiring that such as attended thereon would intrust him with any parcel of that Saint's relics, and if Beggars might be permitted to be choosers, with his hair-girdle which he did wear next his skin; giving commission to *Feliciano* to be bound to what proportion should be required, or to procure security in the City for the restitution of the said relics in ten days; and to send along with it some *Priest* eminent for his devotion, upon assured confidence, that the virtue of the relic meeting with his prayer, should produce some strange effects towards his recovery.

Feliciano is proud of his employment, hoping hereby to ingratiate himself. He makes all possible speed he can to his journey; but first a saint of his must be saluted, and the fair hand of *Paduana* kissed, which done, he sets forth with such alacrity, as resolved, that expedition and faithfulness should contend in him, which of them two should share the greatest part in performances of *Bondi's* commands.

The day after his departure, *Monsieur Insuls*, a

Frenchman, arrives at *Venice*, son and heir to *Monsieur Opulent*, the rich merchant of *Marseilles*. He had purchased three French Counts out of their Lands, beside a vast bank of money in Venice and elsewhere. The old men some months since, by letters, had concluded the match between the two sole children and heirs.

Insuls then comes now, not so much to woo as to wed. Portion and Dowry are both agreed on, and nothing wanting save that, without which all was nothing, the affections of *Paduana*. This *Insuls* was a Poetical fool, an admirer of his own rhymes, rather than verses, being but one degree above *Ballads*.

Yet to give him his due, sometimes he would stumble upon expressions which might have become a wiser and more learned man. And although herein he was generally condemned for a Thief, that he had stolen them from others (his memory being better than his fancy), and then confidently vented them for his own; yet others were so charitable that he wore not the Periwig of other men's wits, but was the true Author of his verses; for he that shoots often at random may sometimes unawares hit the mark, and it is impossible that in a million of blanks one prize should not happen at last. Besides who knows not, that the veriest of fools have not their wise intervals; sometimes he would utter himself in convenient language and quick conceit.

To be brief; it was a great question whether *Paduana* more perfectly hated him, or more entirely loved *Feliciano*. But her Tyrannical Father, driving that affection which he should draw, and forcing what he should persuade, vowed by Saint *Sylvester*, his usual oath, that he would disinherit her of all his estate, and leave the same to *Georgio Bondi* his Nephew, in case she made the least refusal herein.

In this distress *Paduana* makes her addresses to *Gervatio* in this manner.

Gervatio, you cannot but be sensible of reciprocal love between me and *Feliciano;* for though my Father be blind, you enjoy the benefit of your eyes, nor can we pretend to so much secrecy but that one as yourself, constantly with us, will observe smoke in a sigh, and sparkles in an eye which have passed between us. My humble request to you is (and do not Mistresses command when they request), that you would invent some way to free me from the torture of this Clown, fool *Insuls*, and promote my affections where they are bestowed and deserved.

Who would worship the setting Sun when the rising Sun doth court him? My Father's decaying age carries in it a despair of Long life, whilst my tender years promise a longer continuance; build not therefore but on that foundation which in probability appears the firmest. However, I would not disoblige thee from my Father; make your own ends upon him, gain of him what you can, and I will confirm it: and over and above I will assure thee (so far as my condition is capable to give assurance) to gratify your endeavours to a higher proportion than you can expect. Is not the house in the suburbs of *Padua*, where I was born, a pretty pile of building? Do not forty acres of ground impartially embrace it on every side? Is not the Oil in *Lombardy* known to grow there? Know, *Gervatio*, that all these are thine to the reward of thy fidelity: besides I conceal something to myself, intending to surprise thee with that, which in my judgment will be considerable in itself, and worthy of thy acceptance.

Gervatio hereby is made a perfect Convert. He will hold with *Bondi* but run with *Paduana;* he will look towards the *Father*, but will row with the *Daughter;* and professed his future fidelity unto her with such oaths and imprecations as commanded her belief.

It happened at this time, a scurrilous, scandalous Libel, made in verse, was cast in the *Piazza* in *Venice*, other copies scattered in Saint *Mark's*, and other public places of the City. Herein the *Duke* and *Senate* were basely abused, and some small wit being showed in the close thereof (as who cannot be ingenious to abuse), spies are set to apprehend the person, with promises of two hundred Ducats for their service.

Gervatio, whose brains now beat about nothing but *Paduana's* happiness, accosteth *Monsieur Insuls*, who ever was inquisitive of news, asking him what was the tidings of the day. Strange news, saith *Gervatio:* an admirable piece of Poetry (but a little salt and bitter) is found scattered before the *Duke's Palace*, and strange it is, that he who had the wit to make it, had not the wisdom to affix his name to it.

That is no strange thing (saith *Insuls*), for generally the most witty are the most modest. How many hundred nameless copies of mine fly about in *France* and *Italy*, and others perchance claim the credit thereof: it is a passage wherein I have taken special contentment, to see the impudence and ignorance of those who will father the issues of another man's brain.

Gervatio returned, that some hundred Ducats were promised to the Author, and he believed that he should be *Poet Laureate* for the State of *Venice*, and wishing that he any way might be instrumental in discovering the composer thereof; and Monsieur *Insuls*, give me leave to be plain with you, I have a great suspicion (but to recall the word, for suspicion is only for things that are bad, and therefore improper at present), I have a strong surmise that you are the Author thereof. *Insuls* laughed with an affected guiltiness, and said nothing.

And Sir (returned the other), I know you, and

none else, could do the same. First, I compare the style and language is like that wherein last night you courted my mistress; always full, but not swelling, sometimes humble, but not flat, the rhyme good, but not affected, teaching it the true distance thereof, that it must be the vassal, not the Master of the Poet. It is neither *Virgil's* strain, nor *Ovid's*, but both; it comes off with a spring in the close, and commonly the two last verses of the Stanza contain the total sum of the particulars going before; certainly a vast sum of money would be bestowed on him who was the composer thereof.

What talkest thou to me of money? (replied *Insuls*) my Father hath the three best seats in all *Provence*. *Crassuss* and *Crœsus* were both of them beggars unto him. I only take up this vein of Poetry for my recreation, and to confute the common observation, that all Poets are Beggars. I am a rich Poet.

After the exchange of some discourse, *Insuls* showed himself to brook his name, and barely confessed himself to be the Author of the *Poem;* adding withal, that he had made five hundred better in his days. But seeing no one drinks alike of *Helicon* at all times, a constant tenure cannot be expected in wits. I was, saith he, at the penning thereof, not ascending but declining *Parnassus;* and good *Gervatio,* make not conjecture of my parts on such disadvantages, but that I am able to exceed it on the least occasion. I confess, Mustard is nothing worth unless it bite ; I put in little sharpnesses to give it a hogo to the palate of the men of these times.

Out springs two invisible witnesses, whom *Gervatio* had planted within the reach of their words, and presently seize *Insuls,* condemned by his own witness. These informers, the necessary evils in a State, were encouraged in *Venice* by the greatest politicians, conforming themselves to all companies, having a patent to be *knaves,* that they might discover *fools;* seeing no wise men, though dissenting from the present power, will lie at so open a Guard as to carry their hearts in their tongues.

Insuls is presently hurried to prison, and it is strange to see the sudden alteration this accident made upon him. He who at best was but half a man, was now less considerable than a beast, senseless and stupid, scarce able to write his condition to his Father; so that had not some Frenchmen by accident visited him, he had certainly died in prison.

Monsieur Opulent hearing of his son's restraint, posts from *France* to *Venice*, the depth of whose judgment could only fathom profit; he was *Sapiens quoad hoc*, wise only in the point of wealth, so that by long living and great sparing he had accumulated much wealth; but take him out of his own Element of bargaining, he was so simple, that he seemed the true original of his son, as his son the true transcript of his Father.

To *Venice* he comes, and with large gifts buys his son's enlargement: the truth was, his son's simpleness best befriended him, who upon examination appeared incapable to be author of the Libel; and in the judgment of all deserved rather to be whipped for a liar (assuming to himself what was none of his own) than to have any severer punishment inflicted upon him.

Young *Insuls* now at liberty, backed with his

Father's presence, renews his suit to *Paduana's* Jointure, and a good estate are the invincible arguments which *Mellito Bondi* cannot resist. He engageth so far at the matter, that only three days' respite is allowed to his daughter, in which time she must be *Insuls'* wife, or else no heir to her father.

In this juncture of time home comes *Feliciano*, bringing with him the Girdle of St. *Sylvester*, antique for the shape and fashion thereof, as having steel buckles, and a rowel at the end thereof. It was generally believed that this was used by *Sylvester* in the way of disciplining himself; and Father *Adrian*, exemplary in that Convent for sanctity, was sent along with the same, seeing so rich a treasure was not to be trusted in any lay hand.

A solemn entertainment is made in *Mellito's* house, and most of the *Magnificoes* of the State invited thereunto; but this feast had been ushered with three foregoing Fasts kept in the family of *Bondi* and his allies, for the better success and more effectual working of their intended design. After dinner Father *Adrian* mumbles as many prayers (it being well if he understood them himself, as confident that none else did in the room), and then ceremoniously the Girdle is applied to *Mellito*, but especially the rowel thereof (wherein most sacredness was conceived to consist) several times rubbed upon his eyes, to so good purpose, that within three hours he so recovered his sight as to discern and distinguish the faces of all present.

Some of the beholders began to suspect some fraud in the matter, only on this account, because the miracle was not instantly done, but successively and by degrees. Let Drugs, and Herbs, and Minerals, which have a natural virtue placed in them, proceed softly and slowly to effect cures; whereas miracles ride Post, and the same moment which begins, doth perfect extraordinary operations. This I say put jealousy in the heads of some present, to doubt the reality of the cure, and suspect some deceit in the matter.

But they being but few in themselves, were quickly overpowered by the number and gravity of those on the contrary opinion; for some of them argued that the rule is not universally true, that all miracles spring in a moment, seeing some of them have been of slower growth, and the same pace hath not always been observed in miraculous motions; and seeing the effects conduced much to the honour of Saint *Sylvester*, every one was suspected for an *Infidel* that did not presently believe: yea, to doubt the truth thereof, was to discover a little Heretic in his bosom that owned the suspicion.

Presently *Bells* and *Bonfires* proclaim the cure all over the City. Persons flocked from all places to behold this girdle, the making and fashion whereof being out of the road of common Girdles, caught and kept the fancies of fond people; some admiring at the matter of it, being, they said, of a Seal-skin; others at the form, as at the weight and greatness thereof, being almost as big as one could well lift. Hence some inferred Saint *Sylvester* to be of a *Giant*-like proportion, above the standard of other men; others collected the general greatness of men in former ages, complaining of the decay of nature, and diminution of men in these days; but the more wiser sort resolved it upon a point of religion, that the aforesaid girdle was worn by way of penance, not so much to strengthen and adorn, as to load and mortify the wearer thereof.

As for *Bondi*, in a large oration he expressed his thankfulness before the company to his titular Saint, whose speech is here too large to insert; only this by the way, to elaborate, not flowing from him freely on the present emergency, but wrought, studied, and premeditated, which again brought new fuel to their jealousy, which suspected some fraud, as if this had been composed on purpose, and conned by heart beforehand, and so the scene of the design solemnly laid. However their budding suspicion was quickly blasted and beaten down with the general congratulation of all people, so that now his recovery was universally believed, so that this miracle gave a supersede as to all other discourses in *Italy* for a month, and commonly was the third course at all great Tables, whereon the persons present took their repast.

Next day were the Nuptials of *Insuls* and *Paduana* to be solemnized, had not the seasonable interposing of *Gervatio* prevented the matter. Presently, by his appointment, in comes two *Con-*

fidents of *Feliciano's* (both disguised in the habits of Friars), and boldly press into the Parlour and chambers of *Bondi*, looking so scornfully on all accosting them, as if they carried written in their faces a Patent for their own presumption, and knew themselves to be empowered with an authority above control. *Bondi* no sooner recovered himself out of his amazement which seized him at first, but that he resolutely demanded of them the cause of their coming and intrusion at so unseasonable an hour (it being late at night).

They seemed careless to satisfy his demands. a thing beneath them, being employed in matters of higher concernment; and proceed without any interruption to draw up an *Inventory* of the several goods and utensils in his house.

Imprimis, in double gilt plate, 500 ounces. *Item* in white plate, 1200 ounces. *Item* in &c.

Then command they him without delay to surrender the keys of his Chests and Coffers, which the other refused to do, summoning *Gervatio*, *Feliciano*, and all his servants to his assistance, which presently repaired unto him; and though the two former were privy to the plot, yet they so cunningly concealed it in their carriage, that no *Tell-tale* smile or *blab-look* of mirth betrayed the least suspicion of their privity thereunto, but composed their faces, with reduced countenances, speaking much anxiety and suspense, to attend the issue and event of so strange an accident. Then the elder *Friar* began. Dismiss your servants presently, and let them attend in an outward room : it is a favour we have afforded out of the respect to your place, though not deserved by your person, that hitherto we have been tender of your reputation (so far as a crime of this nature was capable thereof), and would not willingly have you sacrifice the small remains of your credit to ignominy and disgrace; we honour the silver crown of age on your head, though it deserved to be placed on better Temples. *Bondi* is surprised with horror and palsy-struck with fear, being guilty to himself of deceit, causeth the room to be voided of all company, and meekly and calmly requesteth them to impart unto him his offence, and their commission.

The other proceedeth. Crimes, though the same in themselves, are not the same when committed by several offenders, but they commence, and take degrees of heinousness from the circumstances of time, place, and person. A concurrence of all these have conspired to aggravate and inflame your guilt; you have a large and plentiful estate, and cannot pretend poverty to yourself (that engine which forceth ingenuous natures to disingenuous actions) prompted you to so unworthy a practice. The Duke *and* State have reflected on you in a great proportion, so that no neglect or discontent received from them could spur you forward to so dishonest a design. Charity itself must turn just against you, and the best rhetoric so far from defending, that it cannot excuse your offence. What? to counterfeit yourself blind, and at the same time to give a lie both to Heaven and Earth, abusing both in one act by an offence, that as former ages will not find an example, so future will scarce afford belief thereof.

But how hard is it to commit one fault, and but one fault: as virtues are not single Stars, but constellations, so vices straggle not alone, but go in companies and grow in clusters. This ground-

platform of your dissembling must have outward outworks and redoubts, to flank, fortify, and defend the same. This mother lie, how fruitful hath it been in a numerous issue of Oaths, and perjuries, as if you intended to scale the Throne of justice with a series of sins, and draw down revenge on yourself. At last, to close and conclude your villainy, you have fathered the same on miraculous recovery, and have abused your titulary Saint, by pretending his relics the immediate cause of your restored sight; but this Saint being rich enough in himself in real miracles, disdaineth the addition of your sophisticated actions, and will not be dishonoured with false honour, which you hypocritically have fastened upon him; yet in detestation of your dissimulation, and to manifest how zealously he disavoweth all falsehood and forgery, he hath been heard at several times in the night in his shrine with a shrill voice to make discovery of your falsehood. And now we expect to obtain from his *Holiness* and this *State* a confiscation of your goods; in order whereunto we are employed by our superiors to take an exact list and account of your estate, both in Lands and movables, till we shall be further informed how the same shall be disposed of.

Bondi looked as pale as ashes, having scarce life enough left to act his limbs with motion; his guilty conscience was not at leisure to enquire into the particulars, but took all for granted, and now expected nothing less but loss of goods and perpetual imprisonment. For that night the *Friars* are contented to repose themselves, and defer the rest of their work till the next morning.

Mean time *Bondi* and *Gervatio* passed a sleepless night, and it would swell a volume to inventory the particulars of their discourse.

Bondi sometimes is silent, and his tears drown his tongue. *Gervatio* desires to make his countenance to attend his Master's in all motions, first readeth and writeth in his face sorrow and silence alternately, as directed by his pattern. At last *Bondi* breaketh forth into the following expressions.

Listen, faithful *Gervatio*, to the Testament of thy dying Master, for I am resolved not to outlive the funeral of my own credit and estate. I behold myself only as a shadow, stripped out of all estate, whereof already I have made the forfeiture; yet it grieveth me not for myself, whose decayed age hath rendered me incapable of much wordly pleasure. It is not considerable with a solvable man who hath it by him to pay a due debt, which will be required a month or two before the exact date wherein it is due. I behold long life, as the playing out of a desperate game at Tables, lost, at the first remove; only it grieveth me for my daughter *Paduana*, whose youth might entitle her to much happiness, and her virtues deserve no less; poor heart, she must now become the scorn and shame of the City, and as an overgrown flower, wither on the stalk whereon she grew, for want of a hand to gather, a Husband to marry her.

Gervatio rejoined, I am utterly unable to give Physic to your other maladies, but possibly may apply a remedy to this, if servants may take the boldness to teach their Masters, and to reprove them too, wherein they conceive them faulty: refuse not an humble reprehension from him, whose good intention and heart may make out what is wanting in his tongue and expressions. I conceive all this misfortune justly befallen you for undervaluing the merits of *Feliciano*, and crossing the affections of your daughter; true it is, his wealth is not considerable, but his extraction, abilities, and accomplishments, doth recompense all his other defects; besides, what loving Parent would stop the affections of his daughter in the full speed and career thereof, except she had bestowed them so unworthily as to entail shame and disgrace on his Family? Now, Sir, make a virtue of necessity, and before the matter be publicly known of the confiscation of your fortunes, comply with her affections, and please him in seeing the marriage between them consummated. You may also by my hands derive unto your daughter so much of your invisible estate as is contrivable in a small room, and may escape the hands of these *Harpies*. It will be safe in all tempests to have a secure place for anchorage, nor can you have any assurance of a better, than with your dear and dutiful daughter. For suppose (and would it were but a supposition, for the case is too plain and pitiful) that all your estate become a prey to their hands, who never let go what once they lay hold upon. Yet I presume your wealth will be a ransom for your life and liberty, and that you may be permitted freely to breathe the Venetian air the short remainder of your days.

Bondi consented to all, as no wonder, for grief had so mollified his heart, it was capable of any impression which bore with it the least probability of comfort; and as a sinking man made an Oak of an Osier, caught at anything to support him from present sinking. The Priest was sent for that night, and though the hour was uncanonical for marriage (long after sun-set), yet the sun of golden *chicqueenes* will make the other sun rise at pleasure; and *Feliciano*, with a largess to the *Priest*, bought off all irregularity.

Then *Gervatio* took the boldness to make another motion. Sir, I humbly conceive that as yet you are not legally convicted, and that there is still inherent in you a power to make over your estate, for the world as yet takes no notice of these clandestine transactions; you are innocent, till such time as you are made to be otherwise by public conviction. I confess myself as unskilled in any science as ignorant in Law; but Law being founded upon reason, methinks I speak proportionable thereunto: beside, my former Master was a chief Advocate, and if my memory, or misapplication thereof, fail not, such fragments of his counsel still remain in my brains, which he often gave to guilty persons in this case before their public condemnation.

Gervatio's counsel passed for oracles with *Bondi*, who in this ecstasy of fear suspected all his own actions, and relied on any man's advice, who would favour him so far, as to bestow it on one so despicable and forlorn as he conceived himself to be. A *notary* is sent for, to make a deed of gift of his estate as fast and firm as his skill would give him leave, and now the same is settled on *Feliciano*.

Feliciano next morning repairs to the two pretended *Friars*, bringing *Bondi* along with him, and desires to know, whether or no it were possible to sopite and suppress the infamy of this action, and to buy off the slander with a round sum of money instantly deposited. The *Friars* confessed the fault great; but because their Convent had been formerly beholden to the bounty of *Bondi*, and because they beheld the fact as of human frailty and infirmity, to which all men are subject, it was hitherto their desire and design to conceal the same; so that, their *Prior and Superior* excepted, none beside themselves were privy thereunto, who gave their attendance when the aforesaid voice made the discovery. They would therefore endeavour their utmost, and nothing should be wanting in them to stop the further proceedings thereof, and doubted not but their pains would take the desired effect, which accordingly in few days was performed.

A Banquet is made, to which many of the *Venetian Magnificoes* were invited, but *Monsieur Opulent*, and *Monsieur Insuls*, his son, because strangers, were by the courtesy of *Italy* preferred to have a prime place among them. There leave we them, feasting themselves with such variety of dainties, that the appetite of many stood long time neuter, not knowing where to fix itself, courted with equality of variety.

As for *Paduana*, it is pity to disturb her any more with our tedious discourse, leaving her in the embraces of her dear and virtuous *Feliciano*, whose name, as it hath in it some tincture of happiness, so took it not its true effect till this time, who was now possessed of a large and rich estate. And *Bondi*, who formerly starved in wealth through the narrowness of his heart, fed better a *Boarder* than a *housekeeper*, having a son and daughter to provide plentifully for him what his covetousness denied to himself, who formerly possessed, but now began to enjoy his estate. Let *Insuls* then return into *France*, and court the *Madames* there, whilst *Paduana* rejoiceth in her choice, and is so far from measuring her promises to *Gervatio* with a restrictive hand, that she outdid his expectations.

FINIS.

TRIANA AND SABINA.

In the City of *Barcelona*, in *Spain*, lived a civil Lawyer of great repute, with a name fitting his profession, *Don Facundo Osorio*, whose office was parallel to a City Recorder in *England*. He had a wife whom he highly affected, and well might she have merited the same, had the Jewel answered the Casket, and her conditions borne proportion to the rest of her corporeal perfections; but there being some disparity in their ages (earth rather than heaven making the marriage between them for worldly respects), her *green* youth answered not his gray hairs with suitable return of affection.

The truth was, she was rather cunning than chaste, and the same was discovered by the friends of her husband; whereof some took the boldness to advertise him, that hereafter he might order her with a stricter hand. But I know not by what fate it cometh to pass, that oftentimes their ears and eyes who have least cause are open to jealousy, whilst those are shut thereunto who have just cause to entertain it. His friends reap nothing but frowns for their faithful counsel. *Facundo* will not believe his wife was otherwise than she should be, measuring her honesty by his own; yet some score this, his good opinion, rather on his policy than charity, knowing what he would not acknowledge, lest it should be a disparagement to his reputation; he saw, but was pleased to wink at his wife's miscarriages; and, because he made the match against the advice of his friends, of his own wilful inclination, he would maintain the ground-work, and own no error therein, lest thereby he should bring his own discretion into question.

One principal friend, *Vejeto*, had his house looking into the garden of Don Nicholayo, a great Lord of that City, who bore unto him no goodwill, because his window hindered his privacy, and was able to tell tales of such passages which he would have transacted in darkness, without any witness. He informs *Facundo* that he had seen some gestures more bold than civil between the Lord and his wife; but *Facundo* still persists in his infidelity, and either believes his wife to be honest, or else acteth the belief thereof so lively, that none could perceive in him the least suspicion of her loyalty.

Sabina, Facundo's wife, falls now seemingly sick, and personateth a dying woman to the life. Her old Nurse, who conveyed intelligence between her and the Lord, had instructed her for her behaviour in a design. Strange it is how she dissembled herein, so that had *Esculapius* himself beheld her (provided he neither felt her pulse, nor consulted with her urine), he might have mistook this his patient to be sick: indeed her husband plies her with Physicians and Physic, all to no purpose, her malady rather increasing, and the fire of her distemper growing the hotter for those cooling juleps which were tendered unto her.

Don Nicholayo repairs unto her, to give her a

visit. Many good counsels he lavisheth upon her in a long and tedious discourse, and the more tedious because her husband was in the presence, and they two not alone by themselves; but at last he recommended unto her a noted *Mountebank*, who had commenced Doctor in the mouths of the vulgar, and had gained to himself much esteem for several palliated cures performed by him; avowing, that if she ever expected health that must be the happy hand to reach it unto her.

This *Quacksalver* had reaped the credit of many learned men's labours, and leaped into esteem by the advantage of their endeavours; for when, by their learned receipts, some able Physicians in *Barcelona* had brought their patients to the point of amendment, and reduced their diseases to terms of easy composition, this fellow would interpose, and insinuating himself into the sick men's acquaintance, would prescribe unto them some of his own medicines, more remarkable for their hard and strange names than any other virtue therein: thus carried he away the credit of many cures, and was cried up by the credulous people for eminency in his art; and although the

Spaniards generally are admirers of themselves, and slighters of strangers, yet this *Mountebank*, being an Italian by nation, had gained among them a great reputation, beholden therein, not so much to his own learning as the others' ignorance.

Seignor Giovanni was his name, who is presently sent for. He comes, views his patient, and after some short discourse, affirms her disease mortal, except one herb procured for her, which grew but in one place of Italy, and must be ceremoniously gathered, by his, or her, hand which bare the truest or deepest affection to the sick body.

Don Facundo her husband desires the *Mountebank* to enlarge himself concerning the name and nature of the aforesaid herb, protesting he would expend the half of his Estate for his wife's whole recovery.

The other, putting on the vizor of a starched countenance by pretended gravity, to procure the greater respect to himself and credit to his words, proceedeth as followeth.

Many men are infected with this singular error,

that they will believe no virtue in drugs further than they themselves are able to render the reason thereof; whereas nature is rich in many secret qualities, which produce occult effects. The herb *Lunaria* may be an instance, which is the greatest picklock in the world; for let it be gathered on Midsummer eve, just at one of the clock, by one looking South-eastward, and some other essential circumstances, locked up in the breasts of artists, it will make any Iron bolt to fly open. The herb *Strellaria* cometh not short thereof in virtue, as useful for those diseases which proceed from hot and dry causes. This groweth but in one part of *Italy*, some leagues from *Lucca*, and I can give infallible directions for the finding thereof. If therefore the Gentlewoman (feigning himself ignorant that she had a husband) had any confiding friend, which will follow my signature in finding and gathering the same (right just at this instant of the year), this, and this alone will restore her to her former health, and I will pawn my credit on the same.

Know by the way, that this *Mountebank* was secretly bribed by *Don Nicholayo* by this design to put her husband to run on one of these two rocks; either to be censured for want of true affection to his wife, if denying to do anything in order to her recovery, or else, with great hazard to undertake a long and tedious voyage, by Sea and Land, to seek her a new nothing; whilst his wife all the while intended to prostitute herself to the amorous embraces of this *Lord*, who had made a mine in her heart, and had supplanted her husband in her affections.

Facundo, that he might be exemplary to all husbands, and that her kindred, who were many and rich, might the more favourably reflect upon him, from whom he had a fair expectancy of a further fortune, willingly undertakes the voyage; desiring to be furnished with perfect instructions from *Giovanni* for the finding of the herb, who delivers him a parchment be-charactered with barbarous figures (Nets, first to raise, and then to catch the fancy of fools) and some other informations, which should be as so many signs and tokens, whereby he should make the hue and cry to attach and apprehend that guilty herb, which having so much worth and virtue in itself, would rather peevishly wither in a private cave than spend itself for the public good.

Facundo was some miles forward on his journey, when *Vejeto*, his former friend, posted after him, and persuaded him to return, for he had urged such unanswerable arguments, and infallible demonstrations drawn from what he himself had seen out of his own window, that at last he prevailed on the belief of *Facundo*, that all was not fair between his wife and *Nicholayo*. Indeed when many scattered circumstances were carefully put together, and seriously perused, there needed no *Œdipus* to read this riddle, which did interpret itself; that *Facundo* must be sent a pilgrimage into *Italy*, to the shrine of an unknown herb (the man in the Moon having eaten *Strellaria* long before), that so, in the vacancy of his bed, the other might be made the incumbent thereof.

At last *Vejeto* adviseth *Facundo* to return home in a disguise, and pretend himself to be a brother of his, long since employed in the Low-Country wars, and now at last, laden with wounds and wants, returned to bequeath his aged body to his native Country. *Facundo* consents, hoping by the well management of this project either to prevent or else to discover his wife's unfaithfulness.

Now whilst *Vejeto* is accommodating his friend *Facundo* with all necessaries (the badges of an old Soldier), all essentials thereunto could not be so conveniently procured but that *Infido, Vejeto's* servant, was employed in completing his provisions, a crafty fellow, who could steal light from a small cranny, and light a candle at a little spark, knowing how thriftily to improve a *small discovery* to the *greatest advantage*. Don *Nicholayo* paid a yearly pension to this *Infido*, to furnish him with intelligence against his Master, who now revealed all the design unto him, for which he received a considerable reward; and *Sabina* is thoroughly instructed to behave herself in the prosecution of the matter. *Facundo* comes to his own house metamorphosed to a *Reformado*, his clothes having so many rents in them, as presumed to cover more wounds under them; a sword by his side which had contracted rust since the last truce. Knocking at the door, his wife sends forth a lamentable shriek, to evidence the continuance of her sickness not dissembled, and *Facundo* (a man of more eloquence than valour) begins to quake, and condemns his own return; but now he was engaged so far, past hope of retreating, that he must either march forward with confidence, or return with shame. Being entered into the house, where he acquaints the servant that he was *Strenuo*, the Brother of *Facundo*, and is conducted by his wife's command to her bedside.

I understand (said she) by my Maid, that you are my Brother *Strenuo*, whom fame long since had reported dead, but we will pardon her that good lie which is better than a truth. I am heartily sorry at your brother's absence, and more, that my indisposition is the cause thereof; never was poor woman more rich in the affections of a loving husband. Though hitherto I have lived his wife, I shall hereafter demean myself as his servant, to deserve some part of his pains in my behalf; no dangers at Land, no tempests at Sea, have deterred him from undertaking a long journey into *Italy*, thence to fetch necessaries for my recovery; but assure yourself (and here she gave so great a groan as broke off the entireness of her discourse, till after the taking of a Cordial she began again)

Pardon, good brother, my unmannerliness in my abrupt discourse; sickness carrieth with it its own dispensatory for such incivilities. I have almost forgotten what I said last, but shall never forget the lasting love of my husband unto me, nor have I any better way to express my affection in his absence, than by using you with all the respect my present condition can afford. You are too noble to measure your welcome by your entertainment, and know full well, servants will not be found where the Mistress is sick. If they fail of my desire, their duty, or your deserts, in their attendance on you, it is in your power either to pardon or punish them, to whose sole disposal I commit the family, and command the keys of all the rooms to be tendered unto you. And now indeed, Sir, the more I look on you the more I love you; your mother never wronged your father I dare boldly affirm, so like you are in stature and complexion unto my husband, that were it not for the difference of your habit I should believe you to be the same.

Alas, Sister (said he), I am two years older than your husband by age, ten years by infirmity : read

wars, I made a solemn vow with myself never to learn or speak a word of that language; for I conceive it a degrading of my tongue to bow to their low expressions, and I admire that any *Spaniard* will offer to dishonour himself by condescending thereunto. Conquerors ought to impose a language on a country, and not to receive it thence: the valiant Romans never learned the *Gaul's* tongue or the Punic language: this consideration hardened me in my resolution, that my tongue should be dumb in *Dutch*, a tongue wherein there is such confluence of consonants, so long, so hard, and so harsh, that it seemeth to me rather made for conjuring than converse, and fitter for Devils than men to discourse therein.

Don Olanzo civilly declined more *Dutch*, and proceeded in his own tongue to sift *Facundo*, desiring him to proceed in the present character of the country.

For Ostend, saith *Facundo*, the only matter of moment, the siege still continueth. We have built three half moons, and a Redoubt, between the windmill and the quay, but the trench from Saint

Dominic's to the counter-gate is not yet perfected. The English out of the town, exerciseth us with daily sallies, and behave themselves very valiantly: the worst can be said of them they are our enemies. Meat beginneth to fail them much, and munition, as we are credibly informed by the fugitives which daily repair to us out of the Town. The Count of *Aremburg* is daily expected with a recruit of two thousand foot, the *Pioneers* out of *Luckland* are daily expected, and then, have at a new mine for the Castle, where all our forces are completed, we shall put it to a desperate assault.

Don Olanzo, not contented with these general heads, wherewith a man might furnish himself out of the weekly news-books, began to press him to the particular description of some places in *Brabant* and *Flanders*. Now though *Facundo* was well book-learned, so far as maps could instruct him, yet was it impossible that they could inform him in all particulars of places and buildings. *Facundo* begins to falter; the other prosecuted him with the cruelty of a prevailing coward, and at length breaking into some choler and passion, caught hold of his beard, which having no better title to his face than glue could afford it, presently fell off, and discovered him to be what he was.

Sabina all this while lay in her bed listening to their discourse, which sometimes she disturbs with her groans and sighs; but now her husband's deceit being laid open, *Facundo*, laden with grief and guiltiness, falleth down on his knees, and craves pardon of his wife.

Strange it is to see how poor spirits descend beneath themselves. But upon his submission and acknowledgment of his fault, a pardon is signed and sealed unto him, upon condition he should resume his journey, which next day with all possible speed he undertook; and we leave him making what speed he might to the place for which he was bound.

The next day *Sabina* falleth truly and really sick. It is ill jesting with edge-tools; that which we play with in sport may wound us in earnest. *Don Nicholayo* repairs unto her, with full intent to enjoy his pleasure, and that nothing but his own moderation should set bounds thereunto, when he meets with an unexpected repulse; *Sabina* complaining of the intolerable torture, which disposed her rather for a Coffin than amorous embraces.

I confess, saith *Don Nicholayo*, were I not privy unto this dissembling, yea the prime procurer and contriver thereof, I should myself verily believe thee really sick. O how far your sex transcends ours in dissimulation: we do it so dully, so improperly, that we are liable to discovery: you exceed yourselves in what you please.

But, Sir, returns *Sabina*, mistake me not. 1 cannot say by the faith of a loyal wife (having to my shame and grief forfeited that title), but by whatever can be true and dear unto me, I vow and protest myself so sick, that nature scarce affordeth me ability to express my own sickness.

You will always be a woman, saith *Nicholayo*, who generally overdo or underdo what they undertake, hardly hitting on a mean, whose souls are either empty or over slow; it is high time now to put off your vizor, and be what you are. And with that he offered a familiar violence unto her, as supposing she expected some acceptable force, to be seemingly pressed to what she desired.

Content yourself, said *Sabina*, or my shrieks shall give an Alarm to the Family. Know, Sir, I never loved you so much as now I loathe your looks, and detest the sight of you. Too much, to my shame and grief, hath formerly passed between us, but now I am resolved not to proceed in that vicious course; but as much as penitency can make a harlot honest, to unstain my soul from my former offences. My time is short, depart the room, and prevent my sending you away.

Don Nicholayo standeth amazed. Who could expect that the wind could blow at such a point of the compass? a cold wind indeed to nip the heat of his lust; yet he seeth no remedy but to comply with the present occasion, and goes home with a soul divided between grief, anger, and wonder; though the latter may seem to claim the greatest share in him at so strange and unexpected an accident. *Sabina* presently dispatcheth a Servant to overtake her husband, requesting him by all *Loves* to return with all possible speed, for she had some important secrets to unbosom to him, nor could quietly depart this world before the imparting thereof unto him.

Facundo, fearing some fraud in the matter, refuseth to return. Indeed the servant accosted him with his Message in that very minute wherein he was taking Ship, the wind serving fair, and most of his goods shipped already. At last the servant gave such assurance of his mistress's sickness, and so importunately pressed him with those arguments borrowed from her, that *Facundo* returns that night home. The room is voided, when *Sabina* begins, with tears in her eyes, moistening her words as she uttered them.

What term shall I call you by? husband, I am ashamed to style you, the mention whereof woundeth my own heart with the memory of my unworthiness; friend is too familiar a title; Lord and Master too terrible to me, a false, deceitful servant; style yourself, Sir, what you please; I am your wretched vassal, and want words to express the foulness of my offence against you. I am ashamed to speak what I blushed not to do, who have lived in a course of inconstancy for many years with Don Nicholayo, and with my dissembling have put you to much trouble and pains. Pardon is too great for me to ask, but not for you to give. I confess they that once have bankrupt their own credit can give no security for the future that they will be responsible to such as trust them. However, Sir, know I place more hope of speeding in my request in your goodness than the Equity of my Petition. If life be lent me, which in my own apprehension (and every one is best sensible of their own condition) is utterly unlikely, I shall utterly deserve some part of your kindness. Sir, can you not see through the chinks of my broken body, my very heart inditing my words; assure yourself —— and there she fainted.

• Servants were called in, and much ado with *Aqua vitæ*. They courted and wooed her soul not to depart her body, which was so sullen that it would hardly be persuaded to stay, though at last prevailed upon.

Although the Passages between *Facundo* and *Sabina* were transacted with all possible secrecy; yet they could not be so privately carried, but that some in the house overhearing it, it crept through the *Family*, and went through the City of *Barcelona*, and at last through the whole Province of *Catalonia*. And now *Vejeto* is found a true friend, and begins to flourish, being formerly so depressed by the greatness of *Nicholayo*, that he took no comfort in himself, and had abandoned his City house, and retired into the country; but now he returns to *Barcelona* again, falls a building and repairing his house, to outface his neighbour *Nicholayo*, making the same both larger beneath with Vaults, and higher above with magnificent superstructures.

Don Nicholayo, enraged in his mind with the discovery which *Sabina* had made, and seeing himself slighted in his reputation, and fearing lest the king of *Spain* (the Court having gotten intelligence thereof) should disseize him of his *Governor's* place of *Barcelona*, which his Ancestors had for three descents possessed, almost to make the honour hereditary, resolves on a design, which present passion prompted unto him, and thus he effected it. Facundo was late at night drawing up some conveyances for Land (which the City exchanged with the Crown) in the Town house, and the employment was certain to engross him until the next morning, which was the last day of the term assigned for the completing thereof. All *Sabina's* servants were by her Nurse (that *Pander* to her former wantonness, and still an active instrument for *Don Nicholayo*) sent upon several errands to places of some distance, and she alone left to attend her Mistress. In springs *Nicholayo* with two robustuous servants, and with violence carries away *Sabina* muffled in carpets (threatening her with present death at the least resistance or noise) to the house of *Nicholayo*. Yet had he now no design of lust upon her, whose revenge had banished his wantonness; and bringing her into his Hall about midnight, a stone there is taken up, and she tumbled down into a vault, which I know not whether to call a *Dungeon* or a *Charnel-house* (many innocents having formerly been despatched in the same place), *Nicholayo* conceiving, that either she would be killed with the fall, or else starved to death in the place.

Now there was in *Barcelona* an Abbot of a

Church, *Jago Domingo*, preferred to that place by *Nicholayo*, rather by the other's favour than his deserts, for guilt had made *Patron* and *Chaplain* mutually great together; they being both often complicated in the same act of baseness, wherein they served each the turn of the other. They were nailed together with necessary secrecy, so that what friendship did act in others, fear acted in them, to contribute their reciprocal assistance in all designs, seeing the lender to-day was the borrower to-morrow. These two plot together, and lay the scene of the ensuing project. At *Matin*-service, when their Convent was singing together in the Abbot's Church, suddenly their harmony was disturbed by an obstreperous voice, which seemed to proceed from a wall above the *Choir*; the voice spake horror, and grief, and pain, shrieking out shrill, and then the noise of rattling of chains and the chinking of Irons was alternately heard. Which indeed was nothing else but an instructed *Novice*, placed there by the *Abbot* in a concealed concavity to play his part, according to his received directions. The Friars for fear, shorten their service, and betake themselves to their Cells in amazement, as utterly ignorant of the fraud; the *Abbot*, his *Novice*, and *Don Nicholayo* being only privy thereunto.

Next day, the Friars meeting at *Matins*, the same noise was heard again, but louder than before, with a clashing and gnashing, speaking and mixture of pain and indignation; the Friars hardly held out their service, wherein wonder so spoiled their devotion, that as at other times the lay People knew not what they said, so now for fear they scarce understood themselves.

The next day was a public Festival, wherein all the Gentry of *Barcelona* met there (save Facundo, who went not out of his house, being transported with grief and amazement what should become of his invisible wife), the spirit proceeding as formerly in shrieks. An *Exorcist* was provided, who by the virtue of holy water and other trinkets, took upon him to catechise and examine the spirit what he was, conjuring it by the power of his spells to answer the truth, and all the truth, at its own peril, if concealing any part thereof.

I am (*said the voice in the wall*) the soul of

Sabina justly tormented in Purgatory, as for my many faults, so chiefly for raising a damnable and notorious lie on *Don Nicholayo*, as if he had conversed dishonestly with me. I had been hurried to a worse place, as having nothing to plead in my own behalf, but that I alleged that this scandal was never raised by my own invention, but put into my mouth by *Vejeto*, and he the Parent, I only the Nurse thereof. And here I am condemned to intolerable torment without all possibility of release, until some signal punishment be laid on *Vejeto*, it being the method of this Court that the accessory cannot be released until the principal first be punished. I am also enjoined to make public confession of my fault, and to request *Don Nicholayo* freely to forgive me, without which my enlargement cannot be procured. And therefore I humbly request the Convent, for sanctity most highly prized, to join with me in my Petition, that that honourable and worthy Lord would be pleased freely to remit my fault herein. I am also to desire you to entreat my husband *Facundo* that he would be pleased to confer on this Convent his Vineyard, lying on the East side of the City, between the gate and the River *Riodulca*, that so by the daily intercession of their suffrages I may be freed from my torture.

The *Exorcist* pressed this she spirit to more particulars, as to know whether her husband *Facundo* did not concur with *Vejeto* to advance the defamation.

The spirit answers, that never woman was happier in a better husband, and that she would not add to her fault and pains to belie him: he was utterly ignorant thereof, and had ever retained a true opinion of her unfaithfulness, had not *Vejeto's* malice rooted it out, with constant and causeless suggestions.

The *Exorcist* proceeded to demand what was become of her body, seeing the report had filled the City that it was nowhere to be found, and a suspicion was raised that her husband or her servants were guilty of conveying it away after they had offered some violence to her person.

Let me not (said she) wilfully heap punishments on myself. I must acquit my husband as altogether innocent, nor can I charge my servants (all whose prayers I humbly desire for the assoiling of my soul) for the least wrong offered unto me; the truth is, an evil spirit violently took me away, both soul and body, that my punishment might be exemplary and unusual, as my fault was above the proportion of common offenders.

The *Exorcist* demanded of her whether any other besides *Vejeto*, had with him joined in that bad advice.

To which she returned that the time allotted for her imprisonment was now expired, being assigned but three hours of three several days, for the begging of the votes of mankind to help her in her extremity; that now she must return as Prisoner, carrying fetters about her to the place of her restraint, never more to appear or be heard more in this world; and with that giving a hideous shriek, and rattling her fetters, she took a sad farewell of the place, leaving all that heard it to admiration.

It is strange to conceive how the female sex of *Barcelona* were affected hereat; all concluded that *Facundo* was bound in honour and conscience to satisfy her request. And some of her kindred brought Abbot *Jago Domingo* unto his house, to move him in conformity to his wife's desire, to settle the *Vineyard* on their Convent as a reward of their meritorious prayers for his wife's enlargement.

Long had the *Abbot* cast a covetous eye upon this *Vineyard* as a little Paradise, for the pleasure and situation thereof: it might for the distance from *Barcelona*, be termed the City in the Country, or the Country in the City. It lay on the side of a *Hill*, which knew its own distance to ascend above the *level*, yet was not overproud to aspire to a barren height. It beheld the rising Sun, which is apprehended the most cordial, when the Virgin beams thereof, uninfected with the vapours of the Earth, first enter on our hemisphere. A wood was in the middle thereof, whereon *Facundo* had bestowed much cost, making *labyrinths* and artificial *mazes*. An *Aviary* also he made therein, stocking it with Birds from all Countries; so that some thought (abating only the *Phœnix*) that the whole kind of Birds, if decayed, might have been recruited hence. A Rivulet called *Riodulca* slided through the midst thereof, and seemed so pleased with the same that, loth to depart from so delicious a place, it purposely lengthened its own journey by fetching many needless *flexures, bendings,* and *windings* therein, as if it intended to show that water could be more wanton than the wood under which it was passed. A banqueting house also was made in the middle thereof, with a *Fountain* and *Statues* of *Marble,* where stones were taught to speak by water-works brought by a device into them.

True it is, many questioned the discretion of *Facundo* in expending so much cost on that which severer folk accounted but a chargeable toy, paying many *pieces of eight* for every pint of *wine* that grew therein; others excused him, that being childless, and having a plentiful estate, this was not only harmless, but a useful evacuation of his wealth, many poor people being used in the making as also in the keeping of this *Vineyard.* But that which most pleaded for his expensiveness herein, and justly endeared him to the place, and the place to him, was that it had been in the name of the *Osorios* three hundred years and upwards; and he had an Evidence in his house not exceeding a span of Parchment in length, and three inches in breadth (so concise was Antiquity in conveying of lands), wherein *John, King of Castile* bestowed this land on *Andrea Osorio* for defending the high Tower in *Barcelona* once against the *French*, and twice against the *Moors*. Pardon *Facundo* therefore if he loved this place or was found thereon, seeing doting on it (the premises considered) was excusable, desiring to transmit this Land, to his *Brother's* Son whom he intended his heir.

Oh, what a brave sweet place would this make for *Abbot Jago's* Convent! Did ever mice bite bad cheese, or were ever *Friars* such fools as to affect base or barren ground? they will be assured of profit or pleasure, or both, wheresoever they fix themselves. The *Abbot* is earnestly set on this place, and will either be possessed thereof, or else the soul of *Facundo's* wife should be left tormented in the place where it was.

Her kindred assault *Facundo* with much importunity to estate the Land on the Convent, which he utterly refused to do, not denying to disburse a competent sum of money, but pardon him if he will not part with his inheritance. But this or nothing else will please Jago, so that they parted in some discontent; yet *Sabina's* friends despair not but in process of time to mould him to the Abbot's desire.

Mean time it would make any honest heart to grieve, though nothing related unto him either in kindred or Country, to see the hard usage of *Vejeto;* how he was hurried to the Jail in most ignominious manner; scarce any in the City so old or so young but would adventure to behold so vicious, so wicked a slanderer (for so he was believed to be) brought to punishment, though his imprisonment was conceived but a preface and preamble to a greater penalty (if escaping with death) which would be imposed upon him. *Vejeto* made all the beholders to wonder, who did read in his face so much spirit and liveliness, as if he triumphed in his sufferings, and rather pitied others than himself in this his condition; yea, his eyes and cheeks had as well mocks as smiles in them, which made the beholders to conceive that besides his own innocence he had some further security, not only to acquit himself, but that his Enemies should come off with shame and disgrace, which made the Jailor to demean himself unto him with the better respect.

Within three days the *Visitor General* of the Order was to come to the Convent, and there in all pomp and solemnity to hear the whole Narration of the matter.

Now let us look a little backward, to acquaint you with the true cause of *Vejeto's* mirth in these troubles, who indeed had sufficient ground thereof. We formerly told how *Vejeto* after his return out of the Country began to beautify and enlarge his house. In sinking a Cellar he stood by the workmen, partly to encourage them with his presence, partly to behold the effects of their discovery, the place being concavous, the ruins of some great structure formerly level with the ground, though now sunk some yards beneath the same. But the general report was, that it had been a Castle in the time of *Julius Cæsar*, when *Barcelona* was a *Roman Colony*, and privileged with municipal communities. *Vejeto* had a great fancy in *Roman Coins*, and would give anything to such as could produce him any variety herein, yet his fancy was not above his judgment, but he was very critical therein, and had an exact and true eye to discover between true and counterfeits, *Casts* and Originals. But, oh, for a *Galba*, whose short reign made his Coins the greatest rarity: and *Vejeto* had a set of Coins with a continual succession of all the *Roman Emperors*, *Galba* excepted.

The workmen find two or three Coins of *Antonius*, the forerunners, as *Vejeto* hoped and expected, of more to ensue; but these poor Souls beheld them as *Æsop's Cock* did the pearl, not knowing how to value them; when *Vejeto*, conceiving these the *Vancouriers* of an Army, and the earnest of a greater payment, though truly not so much out of covetousness as curiosity, dismissed them that night, the night indeed dismissing them (it being late enough to leave work), and enjoining them to return next morning.

All his servants being asleep, he alone with his eldest Son *Speano* turn *Pioneers*, to dig somewhat deep, and to sift the rubbish therein. It happens that they pierced a hollow place (and hollowness being a great friend to the conveying of a sound), they heard a strange noise, too big for a Child, too small for a man; this noise was seconded with some light, but so doubtful and glimmering, that it conquered darkness but one degree. *Vejeto* with his son enter the vault: what should innocence be afraid of? The father was ashamed to betray fear in the presence of his Son, as having more experience, the Son scorned to betray fear in the presence of his Father, having his youthful blood and spirit to support him; together they both so order the matter, that they went into the Vault, putting out their candle which led them, and resolved in darkness and silence to expect the Event of the matter.

Down something tumbled, and presently all light vanished, and they for a time sat still to concoct with themselves the reason of so strange an accident.

Presently they hear a groan, such as speak the Soul, neither friend nor foe, to the Body, but such a distance between both, as if willing to depart. *Vejeto* enjoins his son, as younger and abler to adventure, to bring forth whatever it was, and up it is carried (not knowing as yet how to style it, *Him* or *Her*). On the lighting of a Candle this bulk appears a woman, but much maimed, her right arm and leg being broken.

What difference is there between the same Body in health and in sickness, between the same clothed and naked, when ornaments of art are used unto it, or when it is left to the dressing of nature? *Vejeto* knows no more hereof than Woman, and never remembers that he had seen the face before; yet they omit not what art could do to restore her to life, which succeeded according to their desire; she is conveyed to a bed, and no accommodations are wanting which might tend to the speeding of her recovery. At last he knoweth her for *Sabina*, before she knew herself: wonder not at her ignorance who had passed through so many worlds, it being a greater wonder that she was alive, than that she did not know where she was or what was become of herself: no Physic nor surgery is wanting to restore her to her former health.

True it is, *Vejeto* would not make use of any out of his own doors, for the better concealing of the matter, but his own wife *Olivia*, being excellent above her sex at such performances, indeed she had never read *Hippocrates* or *Galen* in *Greek*, yet was she one who by Kitchen Physic did many and cheap cures to poor people, taking only their thanks (and that only if they were willingly pleased to give it) for the reward. From Physic she proceeded to surgery, and was no less successful therein: this made many to hate her who were of that profession; whilst she cared the less for their hatred, as over-balanced with the love and respect which others (but the poor especially) did bear unto her.

Sabina is privately concealed here for some days, whilst *Vejeto*, as we have formerly mentioned, was carried to Prison, where Olivia daily visited him; and it was the general expectation of the people that forfeiture of his estate was all the mercy justice could afford him.

And now *Abbot Jago* is busy in preparing entertainment for *Padre Antonio, Bishop of Lerma,* and Visitor-General of his order, but especially of this Convent, in *Barcelona*, which three hundred years since was founded by a Bishop of *Lerma*, leaving to his successors the hereditary power of inspection over the same, to add, or to translate Orders and persons therein, as advised by their own discretion. *Padre Antonio* was generally hated by the Friars for his severity and austerity of life, being over rigorous in the observation of the conformity thereof; he would not abate them any point, but confined them to the height of observances. Only this preserved his reputation with religious men, that he used others no worse than himself, *practising* in his own person what he prescribed in others, leaving an example of abstinence to all the Country.

No wonder if the Truants shake when the Schoolmaster cometh among them. *Abbot Jago* is jealous that some flaw will be found in him, to oust him of his place, being conscious to himself of many Enormities; for though he walked by that rule, *if not chastely*, yet *cautiously*, he managed his matters with all possible privacy; yet he suspected that Goldsmiths would not receive that false Coin wherewith common people are deluded, and the sharp judgment of *Antonio* quickly discover that which was invisible to common eyes; wherefore, to mollify him in his visitation, by two Friars he sent him a *present* of rich plate to the value of five hundred *Crusadoes*. *Antonio* refused the acceptance, charging the Friars to return it, adding withal, that if *Jago* had observed the vow of poverty, according to his order, he could not have achieved so great an estate.

Three days after, the Visitor comes to *Barcelona* on foot, unattended save with one servant alone; forth rode the *Abbot* to meet him on his *Mule*, and most of their Convent in like manner, together with the *Officers* of their house, and all *Dependants* of the same, to the number of sixty persons. These expected to have met the Visitor in so solemn an Equipage as had been observed by his *Predecessors*; and finding their expectation confused, *Jago* was much discomposed thereat, and so disturbed in his mind, that he forgot that premeditated Oration which he had artificially penned for the Visitor's entertainment.

Here *Jago* proffered the *Visitor* the convenience of his *Mule*, which he refused, adding, with a stern countenance, that so much pomp and state became not men of his profession; conducted he was into the Convent, where such a supper was provided for him as had made provisions dear in the Town. The *Visitor* commanded the poor people to be called thither, on whom he bestowed all the *cheer*, betaking himself to some mean *Viands*, which he brought with him, and thereon took a sparing refection.

Next day the Court was kept, and several misdemeanours were presented unto him. *Don Nicholayo* complained to the *Visitor* of the high offence of *Vejeto*. For although the power of the *Visitor* extended only to the Convent, yet lately he was empowered with a Commission from the *Conclave*, to take into consideration all business in the City which any way related to the late wonder of the spirit in the wall, and to proceed against all persons concerned therein, as he should see cause for the same.

Vejeto is sent for out of prison, and his fault inflamed to the height by the *Rhetoric* of a young *Advocate*, retained by *Don Nicholayo* to set forth the heinousness of the offence. Being demanded what he could say for himself, he requested that one witness might be produced, and her testimony solemnly taken, which seemed so equal a motion, that it could not any ways be denied.

Presently he bringeth forth *Sabina* by the hand (who stood by, but disguised and concealed), tendering her there to the *Visitor*, to make a narration of the whole story.

The *Visitor* is for a while so taken up with wonder, that his soul was at leisure to do nothing else but admire, to see one reported dead and carried away soul and body, alive and in good health. *Facunio* standing by, requested the *Visitor* to favour him if he transgressed the gravity of that Court by bestowing a salutation on his dear wife; being confident that the strangeness of the accident would sufficiently plead for his presumption therein. Next day the *Visitor* proceeds to censure. First, *Abbot Jago* was expelled his Convent, and condemned to perpetual imprisonment: it is thought it had cost him his life, had not some reverence and respect to his order mitigated the Censure.

The *Novice* in the wall, as yet was but a probationer, and not entered in the orders, was condemned to be publicly whipped in the market place of *Barcelona*, being all the blood that was shed in this Comical story. The *Exorcist* pleaded his own innocence, as not privy to the cheat, and that he only proceeded according to the rules of his own Art, whose plea was accepted.

Don Nicholayo, because a person of great honour, highly descended and allied, was remitted by the *Visitor* to the King's disposal, and sent Prisoner to *Madrid*, where he was ordered to lose his office, and fined *ten thousand Crusadoes* to the *King*, and *five thousand Crusadoes* to *Facundo*. *Vejeto* was deputed to succeed *Nicholayo* in the *Governor's place of Barcelona*. *Facundo*, something to compensate his patience, was promoted to be *Advocate General of Catalonia*. The *Nurse* of *Sabina*, privy to the plot, pleaded the age of seventy, and under the protection thereof, at *Sabina's* humble entreaty, was pardoned. *Infido*, a cheating servant of *Vejeto*, who had oftentimes betrayed his Master's secrets, was branded in the face with F. S. (*false servant*). *Facundo* and *Sabina* lived many years together in Love and Credit, and whereas formerly she was issueless, made her husband afterward happy with a numerous posterity.

FINIS.

ORNITHOLOGIE;

OR,

THE SPEECH OF BIRDS.

BY

THE REV. T. FULLER, B.D.

LONDON:
PRINTED FOR JOHN STAFFORD, AND ARE TO BE SOLD
AT HIS HOUSE, AT THE GEORGE AT
FLEETBRIDGE. 1655.

TO THE WORSHIPFUL ROGER LE STRANGE,
ESQUIRE.

Sir,

A most learned Dutch *writer* hath maintained *that birds do speak and converse one with another;* nor doth it follow, that they cannot speak, because we cannot hear, or that they want language, because we want understanding. Be this true or false, in *Mythology* Birds are allowed to speak, and to teach men too. We know that a man cannot read a wiser, nor a child a plainer Book than *Æsop's Fables.*

These *Birds* now come to make their nest under the *Boughs* and *Branches* of your Favour. Be you pleased, Sir, to extend your shadow over them, and as they shall receive succour from you, you may be assured you shall receive no hurt from them.

And thus, Sir, I wish you all happiness, not only to converse with birds in the lowest Region of the Air, sometimes styled Heaven, but that a better and higher place may be reserved for your entertainment.

J. S.

ORNITHOLOGIE;

or,

THE SPEECH OF BIRDS.

———

THERE was a *Grove* in *Sicily*, not far from *Syracuse*, wherein the *Greek* and *Latin Poets* had made many *Hyperbolical* descriptions. For the *Wits* in that Country being *placebound*, and confined to a narrow Circle of ground, sought to improve the same by their active Wit; whereby they enlarged every *Ditch* into a *River*, every *Pond* into a *Lake*, every *Grove* into a *Forest*, every convenient Hill into a *Mountain*. In this notion they magnified this *Grove*, otherwise not above twelve Acres of ground, though well wooded, save that the tyrannical Oaks, with their constant *dropping*, hindered the *underwood* from *prospering* within the compass thereof.

There was the whole Nation of *Birds* living under the shadow thereof. And the climate being indifferently moderate, and moderately middle, wherein the *East, West, North,* and *South* of the *World* were in some kinds compounded, *Birds* of all Climates here made their habitations. Now a Bill of complaint was subscribed (or rather signed) with the numberless *Claws* (instead of hands) of *Birds*, containing the many insupportable Grievances they had endured from the intolerable cruelty of the *Eagle;* who making his own *lust* his Law, had domineered over all the *winged Nation.* The *Eagle* appeared in answer hereunto (it being a general meeting of all kinds of *Birds*), and endeavoured to justify his proceedings, and clear himself in vain from their Accusations. The truth is, the *Eagle* was overgrown with Age, for he is generally reported the survivor of all *Birds.* So that if one would take a *Lease* of Land on a bird's life, he could not put in a more advantageous name than an *Eagle.*

But this *Eagle* had its *bill* with long age so reflexed back again into his mouth, that he was so far from preying on another, that he could not swallow any *Flesh* though proffered unto him. Soon will the *spirits* fail where the belly is not fed; in vain did his *courage* pretend to his wonted *valour,* when there was nothing within to justify and make good the offers thereof. So that the poor *Eagle,* conquered rather with its own *Age* than outward *violence,* yielded to that to which all must yield, and was forced patiently to digest all affronts offered unto him, and glad so to escape. For although some mention was made of killing him, yet by plurality of suffrages that vote got the mastery which only confined him to a neighbouring *wood,* on condition that death without mercy should be his penalty if exceeding the Bounds thereof.

This done, Proclamation was made three days after, that the whole *Species* of *Birds* should appear for the election of a *Principal* to command them. Indeed there were many which were altogether against any Government, because they might the more freely rove and range in their Affections. These held that all were free by nature; and that it was an assault on the Liberty of man, and a *rape* offered to his natural freedom, that any should assume authority above another. These maintained (what certainly was not only a paradox, but a flat falsehood) that nature at the first creation made all the world a flat level and *Champ*, and that it was by the violence of the *Deluge* or great Flood, which by the partial fall or running thereof, made the inequality, by sinking some places into humble *Valleys*, and swelling others into aspiring *Mountains*. Prosecuting which comparison, they maintained that all men were naturally equal; and that it was the inundation and influx of human *Tyranny* which made this disparity between them. They also defended the argument, that as the world began, so it should end with the Golden age; and that all ought to be restored to that primitive Liberty which men had lost, partly *surrendering* it by their own folly and easy nature, partly *surprised* into their own slavery by the cunning and craft of others that practised on their simplicity. But however that these made a great noise, the opposite party prevailed, as having most of strength and reason on their side.

For where all rule, there no rule at all will be; where every man may command, in fine, none will obey the dictates of his own reason, but be a very *vassal* to his *passion*. Society cannot be twisted together where there is not a subordination and subjection one to another; and where every one is absolute in himself there is an impossibility of any orderly subsistence.

Let the maintainers of the contrary try with themselves to make a rope of sand; where each crumb therein being independent of itself, hath no tendency to a general agreement, but enjoys itself in its own entireness.

It being now cast (by general suffrages) for a *Commander* over·all, that at such a time they should meet; it was also proclaimed that all antipathy should cease between all *Birds* during their meeting; because being now in danger of general ruin for want of a *head*, all private animosity should be broken off and drowned in a public agreement.

According to the Proclamation, they all met together; and *birds* of all feathers had a general *convolancy*. Then the Ostrich began, in a high commendation of himself, how he was the biggest of all *birds*, and therefore the fittest to be their

General, as of the greatest ability to support the weight of the *Massive* affairs of *State*. The rest of the birds gave him the *hearing*, until the little *Wren* thus returned the answer.

It may seem a very unproportionable combat between the *least* and *greatest* of *birds*, that I should once offer to enter the *lists* with the *Giant*, who affrights us all with his greatness. But sure this wise *Senate* never made the bulk of a *body* the standard whereby to measure the perfections of the mind: and therefore I may take to myself the confidence to examine the truth of what he hath spoken. His greatness is apparent to every eye; but as for any other eminency, it is so secret a quality, that none as yet hath discovered it. For mine own part I conceive him rather *beast* than *bird*, and therefore not properly of our kind. I appeal to his *Latin* name, *Struthiocamelus*, wherein the *Camel* bears away the last and best part thereof. And are we put to such a strait, that we must elect an *Hermaphrodite*, a rudiment, which is a measuring case between *Beast* and *Fowl*. Doth he not more trust unto his *Legs* to *Flee* than unto his *Wings* to *fly?* and what, I pray, is the remarkable virtue which commends him to public notice? Hath he any *melodious voice* whereby to charm the attentions of those that hear him? hath he any extraordinary *wit*, in which he appears above others of the same society? What if Foolish women, as light perhaps as the *Feather* they wear, be pleased to advance his *Tail* above their *heads*. What if vainglorious Captains, more known by their *Plumes* than their *performances*, deck their *crests* with the spoil of his *wings:* all these amount not to argue any real worth in him. We live not in an age to be deluded with shows, or cheated with shadows. It is enough that our *Ancestors* have suffered for their folly herein, with their own credulity. Real worth must be the attraction of our *love* and *respect;* which being here wanting, I utterly disavow his Election for our *Sovereign*.

The rest of the *birds* concurred with the resolution of the Wren, highly applauding it for the same, which durst *speak* that which others thought. They plainly saw that *spirit* united in a small bulk acts most vigorously; and the contracted heat in so small a *body* prompted the *Wren* to

s

such lively expressions which bigger *birds* durst not utter.

Next stood forth the *Parrot*, insisting largely on its own commendation, among the rest, of his dexterous faculty in imitating the speech of Man, wherein he exceeded all other creatures in the world. And seeing man was the Sovereign of all the Creation, he conceived himself (which approached next unto him in his happy expressions) deservedly might claim the *Regiment* of all *birds*.

The *Daw*, generally condemned for its *loquacity*, took upon him to answer the *Parrot*. Indeed he began with great disadvantage, none expecting anything of wit or worth from him, because he was so common a *Talker*, therefore conceived his *speech* not worthy their attentions; when defeating their expectations, and deceiving them with a harmless cheat, he thus proceeded.

You have heard the *Parrot* make a large *encomium* of himself, all which must needs be true, because you have heard his own credit to avouch it; otherwise methinks one might justly take the liberty to examine the ground of what he hath spoken. I will not insist on the *alienness* of his *extraction;* we living here in *Syracuse,* whilst the *Parrot* fetcheth his Original from the *South* of *Africa* or *East* of *Asia.* Only consider with yourselves how unfit it is for our free-born spirits to submit to a Foreigner. Assure yourselves foreign *Air* will bring in foreign *inclinations:* he cannot but promote strangers as his favourites to all places and preferments of profit and honour; and can this be digested by such as consult the true spirit of an ingenuous Birth? For mine own part, I shall rather submit to the tyranny of our own Country than to the insulting humours of strangers; as expecting that although one of our own Country may for a time domineer over us, yet the *sympathy* of blood to those of his own Land will give a *check,* and at last gain a *Conquest* of his *passion,* that he will return to a favourable reflection on those who by *vicinity* of *birth* and *breeding* are related unto him.

Now whereas the *Parrot* boasteth that he doth so exactly imitate the speech of *Man,* it affecteth me no whit at all with admiration thereof. I have heard of a speech of *Alexander,* who being invited to hear a man that sung like a *Night-*

ingale, answered, I scorn to hear him, for I have heard the *Nightingale* itself; and who would admire the *Copy* when he hath the *Original?* I have often heard men themselves speak, and therefore am not a whit moved to hear a *Parrot* speak like a man. Let everything appear in its own shape, *Men* speak the language of *Men, Birds* of *Birds*. Hypocrisy is that which hath betrayed the world to a general *delusion*, thence to *destruction*, when people counterfeit the *Tongues* and *Tones* of those from whose *Hearts* they dissent. How many demure people hath this age brought forth, sadly and soberly dropping forth their words, with much affected deliberation, as if all the hearers were bound thereby to believe them as solid, reserved, and discreet in *Deeds* as in their *words*, when they only *Palliate* and cloak a base and unworthy *inside* under the shadow and pretence of an *outward* fair representation. I therefore must throw my *grains* into the *Negative* scale, and conceive the *Parrot* utterly unfit for the sovereignty of *Birds*.

After many debates and disputes, pro and con, plurality of voices at last pitched on the Hawk, as whose extraction was known to be honourable, valour undoubted, providence or forethought admirable, as appeared in the quickness of his eyes, being a *Prometheus* indeed, foreseeing all dangers, and his own advantages of great distance. The Hawk returning his full and fair thanks unto them for their free favour, accepted of their proffer, and all their meeting for the present was diminished; only two birds commanded to stay behind, the Phœnix and the Turtle Dove, whom the Hawk severally accosted, beginning with the former.

Sir, or Mistress Phœnix, saith the Hawk, for I know not in what Gender to address my language unto you, in whom both sexes are jumbled together. I desire to be informed of you, whether that be a truth, or a long-lived common Error, of the manner of your original from the Ashes of your Ancestors. If it be a truth, I stand ready with admiration to embrace and entertain it. If an error, I am resolved Posterity shall no longer be deluded therewith. We live in an Age of Knowledge, the Beams whereof have dispelled those mists of Errors wherewith our Forefathers were cheated into the belief of many impossibilities recommended unto them by Tradition, as if the gray Periwig of Old age should command so much veneration from us, that we should consign up our judgment to the implicit belief of anything which former Ages have related. Deal, therefore, openly with me, and inform me the truth, whether your Generation be thus by Continuation of a Miracle.

I cannot resolve you herein, saith the Phœnix, of the particulars of my Extraction, which happened long before the register of my memory. Sure I am there are no other of my Kind for me to couple with, which demonstrates the truth of that which is generally received. I confess men make use of me rather for a Moral, and an Emblem to denote those things which are rare, and seldom come to pass. Thus a Court Lord who will honestly pay all his Debts is accounted a Phœnix: a judge who will not suffer his Conscience to be robbed by a bribe secretly proffered unto him is a Phœnix: a great man who looks straight forward to the Public good, not bound on either side with his own interest, is a Phœnix. However, assure yourself, that besides the Morality that may be made thereof, I have, as you see, a real Existence in Nature; and if any will take the pains to travel into *Arabia*, to *Mecca*, he shall find my Nest in a Tree, hanging there almost as Artificially as doth the Tomb of *Mahomet*, bribed by an invisible Loadstone into that miraculous posture thereof.

But now, saith the Hawk, suppose I should seize on you this night for my supper; whether do you think that the loss of your life would be so great a defect in Nature that the whole *Universe* would fare the worse for the same?

Undoubtedly it would, saith the Phœnix, for this is received for an undoubted Maxim amongst *Philosophers*, that if one whole Kind or *species* of Creatures be destroyed, the whole world would be ruined thereby. For every kind of Creatures are so Essential to the well-being thereof, that if any one of them be utterly destroyed, all the rest of sympathy will decay.

I conceive not, saith the Hawk, that you are such a foundation-stone in Nature's building, that the taking you away will hazard the whole Architecture thereof. However, I am resolved to put it to the trial, be it but to gain knowledge by the experiment. I know what *Plato* saith. *That those are the happiest kingdoms wherein either their Kings are Philosophers, or their Philosophers their Kings.* Seeing therefore the History of Nature is so necessary to an accomplished Governor, I, who desire all perfections in that kind, will, to satisfy my curiosity, make proof thereof.

The Phœnix pleaded for herself the benefit of a Proclamation of liberty to all for three days, to come and go with safety; the Hawk smiling at her silly plea, informing her that such grants are to be kept no farther than they are consistent with the conveniency of those that grant them. Yet for the present the Phœnix was reprieved, because the Hawk's stomach, lately gorged, had not as yet recovered his appetite to his supper.

Then the Hawk approached to the Turtle Dove, demanding of her whether it was true or no what passeth for a common truth, that the Turtle, if once losing their Mate, never wed more, but pass the remainder of their doleful days in constant widowhood.

Most true it is, saith the Turtle, which I may speak by my own sad experience; for some three years since the unhappy shot of a cruel Falconer deprived me of my dear Husband, since which time I have sequestered myself from all company, never appearing in public till now, forced thereunto by command from Authority.

And surely, I conceive all second Marriages little better than excusable lust; for when once the heat of youth hath been abated in one Match, none can pretend Necessity of Marrying again, except it be for quenching those heats which they themselves willingly and wilfully kindle. Besides, when one hath once really affected a Husband, or he a Wife, affections so engross the whole soul, that notwithstanding his or her death, it can never admit another to the same degree of dearness. Especially if their love were signed and sealed with Issue, as mine was, having three of both Sexes surviving (send them better success than their unhappy Father had), so that in them methinks I behold my Husband still alive. She therefore that hath not the modesty to die the Relict of one man, will charge through the whole Army of Husbands, if occasion were offered, before her love will meet with a full stop thereof.

You are too rigid and severe, saith the Hawk, to make your personal temper and private practice the rule to measure all other by, unacquainted with the Necessities of others in this Kind. But to come closer to the matter, I desire satisfaction in another thing, namely, whether you be without a Gall, as is commonly reported.

I know there is a twofold knowledge, one by the fruits and the effects, which Scholars call *à posteriori*, and this is the more fallible and uncertain; the other *à priori*, from the Causes, and this, as more demonstrative, may safely be relied on. I will embrace the latter course, and to assure myself whether you have a Gall or no, I mean to make you a living Anatomy, and instantly to dissect you. Ocular inspection is the best direction, and I will presently pry into your entrails for my better information, to see with what curiosity Nature hath contrived the things therein, and how many little engines there are, to move the wheels of life within you.

Then began the Hawk to dispose himself for Supper, intending the Turtle Dove for the first course to begin with, and the Phœnix (as the finer flesh) to close his stomach therewith. In preparation whereunto he plumed the Dove of some of her upper Feathers.

Just in the instant as he began his prey, who should come in (but he was little expected and less welcome) to the Hawk, than the old Eagle; and we must a while dwell upon the cause and manner of his enlargement.

This Eagle was, as aforesaid, confined to a Grove, where he was temperate against his will, as not being able to feed on any Fowl. Nature had hung such a Lock upon his Bill, for the Redundancy thereof was such, that he was capable of no food save drink, which he plentifully poured in; thus for some months drink was all the meat he took, which served to support his life, though not to increase his strength; yet could he not be a good fellow in his Cups, as being solitary by himself, having none to keep him company.

At last he descried a sharp Rock, wherein one place white in colour, more prominent than the rest, had a shining hardness therein. To this the Eagle applies his Bill, and never left off rubbing, grating, and whetting his beak thereon, until at last he quite whetted off the superfluous, yea hurtful Excrescency of his Bill, which, now reduced to a moderate proportion, was as useful to all purposes as ever before. Thus enabled to get his prey, in a few weeks he recruited his strength, so that what the Poets tell of *Medea*, that with her enchanted Baths made her Father-in-Law young again, here truly came to pass. And now the New old Eagle, hearing in what Quarters the Hawk kept his constant residence, thought on a sudden to have surprised him, had not the other discovered his approach, and made a seasonable escape, whereby both Turtle and Phœnix obtained their liberty, and securely returned unto their own Nests.

The Hawk having made an escape, posted with all speed to the Lapwing, which with some difficulty he found out, and privacy being obtained, thus kindly spake unto him.

Friend Lapwing, I have taken notice that you are the most subtle and politic Bird in all our Commonwealth; you have the art so to cover your intentions, that they are not obvious to common eyes. When your Eggs or young ones be a mile at distance, you use to flutter with your wings, and fetch your rounds and circles a great way off, as if you intended to brood that place with your wings, or as if that were the Chest wherein your Treasure was deposited; this makes many people to search there for your young ones, but are frustrated of their hopes, you having secured them far off. This lawful Simulation I conceive a commendable and necessary quality in every great person; it is as necessary as breath to their well-being. Should men play all above board, and expose their actions to all Spectators, Folly and Wisdom would be both of a rate. No, it is the hanging of such Curtains and Traverses before our Deeds which keep up our Reputation, and enable us for great performances. Now I request you to help me a little in my extremity: the renewed Eagle is in pursuit of me, and my safety lieth much at your disposal. The Lapwing promised the utmost of his endeavours, and desired the Hawk to proceed.

See you, saith the Hawk, yonder empty cage of great receipt, so that it might serve for an Aviary, for which it was first intended, though since disused. When the Eagle, flying this way, inquireth after me, persuade him I am flown into the cage, and leave the rest to my performance.

All was acted accordingly : the Eagle demanded what was become of the Hawk. The Lapwing returned, *Here 'tis, here 'tis*, and then hovered over the Cage, fetching so many compasses thereabouts, that one might have mistaken him for some Conjuror, making his many circles with intent to raise up some spirit thereabouts.

The Eagle violently flies into the Cage, whose doors stood open, triumphing in his own happiness, that now he should be revenged on his professed Enemy. Instantly the Hawk (who stood behind unseen in a place of advantage) claps an Iron Padlock on the Cage, and thus insulteth over the Prisoner.

Methinks, Sir Eagle, you make me call to mind the condition of *Bajazet*, the Great Turk, whom *Tamerlane* took captive, and carried him about the Country, that all people might feed their gazing eyes upon him; such a spectacle are you this day. I have now made an Owl of the Eagle, turned him into the ridiculous object of laughter and contempt. Tell me, do you not want a *Prometheus*, to feed upon his fruitful entrails, as the Poets feign, which daily increased, and afforded the Poet's Eagle both Commons and Festivals? Sir, your life shall not be vented out at once, but you shall die many deaths, with long lingering torments. I will order it so, that you shall feel yourself to die. There is no Music in an Enemy's death which is not accompanied with torment; and though no outward torture shall be inflicted upon you, yet know, that thirst and hunger shall be your two Executioners. Now the Guiltless blood of so many Birds and innocent Lambs and hurtless Hares shall be required of you; and so I leave you till to-morrow, when I mean to make a new meal of you in scorn and contempt.

The Eagle sadly, yet stoutly answered, my courage shall not abate with my condition, whose spirit is planted above the battery of Fortune. I will never be less than myself, whatsoever befalls me. A Lion is no less a Lion though in a grate. Mischance may make me miserable, it shall not make me base; I will bear my troubles with as much cheerfulness as I may, 1 defy thy spleen in triumphing over me.

After the Hawk's departure, the Ostrich came in the place, whom the Eagle saw unseen, and wistily marked his postures and motions. The Ostrich fell into a strange passion, and would you know the reason thereof, it was as followeth.

Some three days since, when he first repaired to the general meeting of the Birds, he left his Eggs in the sand, not covering them over, such his carelessness and incogitancy; it was in a Starlight night, wherein he took a mark for the finding of his Eggs by such a star, under the direct position whereof he then hid them, and hoped to find them at his return. It happened that the Star being turned about with the circumgyration of the heavens, which continue in constant motion, the Ostrich lost the Star by which he thought to find his eggs, and though very near the place, wandered up and down, and could not light upon it, which made him break forth into this passionate complaint.

I am the most unfortunate of all Fowls. How will all condemn me for an unnatural Parent, who have been thus careless of mine own issue? Yet I took as good notice of the place as I could. All things in Earth are false, and fading, and flitting away. I had thought there had been more faithfulness in the Heavens, more assurance in the Skies. Let never the *Indians* worship Stars again, when they are guilty of so much deceit.

How comes it to pass that the Pole-Star is so perfect a guide and direction to the Mariner, that it may be termed the grand Pilot of all Ships, by the Elevation, or Depression whereof, they infallibly collect in the darkest nights whereabouts they steer? I say, how comes that Star to be so true to its trust, to be so true a Conductor of wandering Sailors, and this prove so false to me? And now will Posterity brand for unnaturalness, who have exposed my Eggs to such danger, though therein all caution was used by me to the height of my discretion. More would she have spoken when grief silenced her; for as those Rivers are shallow which make a noise, whilst the deepest streams are tongue-tied, so those passions which vent themselves in words discover their bottom of no great depth.

Meantime the Eagle looked through the spaces, or intervals in the Cage, and so excellent the sight thereof, he easily discerned where the Eggs lay, the Ostrich being so near, that he almost crushed them with his own feet: wherefore calling the Ostrich unto him, I am glad, saith he, that in my misery I have the occasion to oblige any. I can tell you where the Treasure is that you seek for, and presently directed him to the same.

The Ostrich was not so overjoyed with its own happiness· but that he bethought himself how to return proportionable thanks to the Eagle; in order whereunto he set his Bill against the Iron Padlock of the Cage, and according to the voraciousness of his stomach, quite devoured the same.

Let privy Councillors of Nature enter into this deep Discourse, how it is possible for such a solid and substantial thing as Iron is, to become food to a Fowl: let them, I say, beat their brains about this Question, harder than Iron; and if they find the true reason thereof I shall prefer their

Ingenuity as stronger than the stomach of an Ostrich; mean time we will be content to rest in the vulgar report, and are satisfied to admire what we cannot understand in such cases, wherein surely there are some hidden and occult qualities, too deep for men to dive into; and these betray a surly and base disposition, which will believe nothing (though Authentically attested by never so many witnesses) whereof they are unable to render the true reason, as if Nature could do nothing but what she gives them an account of how she doth it.

The Eagle thus restored to liberty, returned hearty thanks to the Ostrich. You see, saith he, there is no living in this world without bartering and exchanging of Courtesies one to another; he that lendeth to-day may borrow to-morrow; how happy would Mankind be if the Wall of Envy were plucked down betwixt them, and their parts so laid in common, that the wealth of one might supply the wants of another. Nature hath enriched me with a quick Sight, thee with a strong Digestion: I have restored thy Eggs to thee, you have restored me to myself, liberty being the life of life; and this I thought fit to testify unto thee, though hot in the pursuit of my Enemy, first to thank thee, then punish him. I will not be guilty of so preposterous a Soul, that my Revenge shall get the speed of my Gratitude.

This done, the Eagle, in full Quest of the Hawk, discovereth a company of Birds together, being a great party whom the Peacock had assembled, with hope to entice them to choose him their Chief; for the Hawk nowhere appearing, and the enlargement of the Eagle being unknown, he thus endeavoured by his Rhetorical flourishes to make himself popular in their affections.

I am not ignorant that such men proclaim their own weakness who are the Herald of their own praise; it argueth a great dearth of desert and want of worth, when one is large in his own commendation. However, sometimes necessity makes it lawful, especially when what is spoken is so generally known, that it commandeth the way to its own belief, and carrieth the credit about it. Give me leave to present my person and merits to your consideration; my bulk not so great as the Ostrich, like to be a burden to itself, yet not so little as any way to invite neglect. A good presence is requisite in a Commander, otherwise, great parts crowded in a despicable person no whit becomes one in Authority. I will give you but one argument, or demonstration rather, of my Worth. When the Gods had the free choice of all the Birds which they would please to make their attendants in ordinary, and when *Jove* made choice of the Eagle, as most Imperial, *Juno* his consort was pleased to elect me, to be called by the name of her Bird in all passages of Poetry. Thus am I next to the best, and but one step removed from the Top, even by those infallible judgments.

Look, I pray, upon my Train, how it is circular, the most capable Form, and how it is distinguished with variety of Colours, which appeareth as so many earthly Rainbows in my Feathers. *Ovid* hath reported that *Argus's* hundred eyes were turned into them. But know you, if you please to elect me to be your Chief, that all those eyes shall daily and hourly watch and ward for your good. I will have a constant oversight of your welfare.

It was conceived that the Peacock intended a longer Oration, which would have wearied the assembly with the Prolixity thereof, had he not casually, but happily cast down his eyes on his black legs, the ugly hue whereof so abated his Pride, that it put a period to his Harangue before his intent and other's expectation. Now as the Vulture was tuning his tongue to return an answer, in cometh the Eagle, and is generally received with all joyful acclamations.

Now because Clemency is the badge of a generous nature, and those that have most courage have least cruelty, at the mediation of some potent Birds, the Eagle condescended that the day of his Re-inauguration should not be stained with blood, and therefore granted life to the Hawk, but on condition not to exceed the Grove in which formerly himself was imprisoned.

FINIS.

ANTHEOLOGIA;

OR,

THE SPEECH OF FLOWERS.

BY

THE REV. T. FULLER, B.D.

LONDON:
PRINTED FOR JOHN STAFFORD, AND ARE TO BE SOLD
AT HIS HOUSE, AT THE GEORGE AT
FLEET-BRIDGE. 1655.

TO MY MUCH HONOURED FRIEND,
WILLIAM STAFFORD, ESQUIRE, MERCHANT OF
BRISTOL.

WORTHY SIR,

In this plundering age, wherein the studies of so many have been ransacked, and many papers intended for private solace and contentment have been exposed to public view, it was my fortune to light on the ensuing discourse. It seemed to me pity that it should be strangled in obscurity, as conceiving it might conduce something to the delight of the Reader; for surely no ingenuous person can be so constantly serious, yea surly and Critical, but to allow some intervals of refreshment not only as lawful but necessary.

Let such morose, yea mischievous spirits, pine themselves to walking Anatomies, who brand all refection of the mind by ludicrous intermissions to be unlawful. To spare a heavier censure (which many more resent of anger), the worst I wish them is always to eat their meat without sauce, and let them try whether their palate will be pleased with the gusto thereof.

In the following discourse there is nothing presented but sweet Flowers and herbs. I could wish it had been in the summer time, when the heat of the Sun might have improved their fragrance to the greatest advantage, and rendered them more acceptable to the smell of the Reader, being now sadly sensible that Autumn, the Usher of Winter, will abate of their sweetness, and present them much to their loss.

Sure I am, no bitter Colloquintida appeareth in this our Herbal; 1 mean no tart and toothed reflections on any. Dull are those wits which cannot make some smile except they make others cry, having no way to work a delight and complacency in the Reader, save only by gashing, wounding, and abusing the credits of others.

It is desired that this discourse may but find as much candidness as it brings, and be entertained according to his own innocency. I have heard a story of an envious man, who had no other way to be revenged of his Neighbour, who abounded with store of Beehives, than by poisoning all the Flowers in his own Garden, wherein his Neighbour's Bees took their constant repast, which infection caused a general mortality in all the winged cattle of his Neighbour.

I hope none have so splenetic a design against this my harmless Treatise, as to envenom my flowers with pestilent and unintended interpretations, as if anything more than flowers were meant in the flowers, or as if they had so deep a root under ground, that men must mine to understand some concealed and profound mystery therein. Surely this Mythology is a Cabinet which needeth no key to unlock it; the lid or cover lieth open.

Let me entreat you, Sir, to put your hand into this Cabinet, and after therein you find what may please or content you, the same will be as much contentment unto your

True Friend,
J. S.

… ANTHEOLOGIA;

OR,

THE SPEECH OF FLOWERS.

THERE was a place in *Thessaly* (and I am sorry to say there *was* a place in *Thessaly*, for though the place be there still, yet it is not itself. The *bones* thereof remain, not the *Flesh* and *Colour;* the standards of *Hills* and *Rivers*, not the Ornaments of *Woods, Bowers, Groves,* and *Banqueting-houses*. These long since are defaced by the *Turks*, whose barbarous natures wage war with civility itself, and take a delight to make a *Wilderness* where before their conquest they found a *Paradise*).

This place is some five miles in length, and though the breadth be coequal with the length, to equalize the same, and may so seem at the first sight, yet it falleth short upon exact examination, as extending but to four Miles. This place was by the Poets called *Tempe,* as the Abridgment of Earthly happiness, showing that in *shorthand* which the whole world presented in a *larger character:* no earthly pleasure was elsewhere afforded, but here it might be found in the height thereof.

Within this Circuit of ground there is still extant, by the rare preservation of the owner, a small Scantling of some three Acres, which I might call the Tempe of Tempe, and re-epitomized the delicacies of all the rest. It was divided into a *Garden,* in the *upper* Part whereof *Flowers* did grow, in the *lower, Herbs,* and those of all sorts and kinds. And now in Spring time earth did put on her new clothes, though had some cunning *Herald* beheld the same, he would have condemned her *Coat* to have been of no ancient *bearing,* it was so overcharged with variety of *Colours.*

For there was *yellow Marigold, Wallflowers, Auriculas, Gold Knobs,* and abundance of other nameless *Flowers,* which would pose a *Nomenclator* to call them by their distinct denominations.

There was *White*, the *Daisy*, *white roses*, *Lilies*, &c., *Blue Violet*, *Irises*, *Red Roses*, Peonies, &c. The whole field was *vert* or *green*, and all colours were present save *sable*, as too sad and doleful for so merry a meeting; all the Children of *Flora* being summoned there, to make their appearance at a great solemnity.

Nor was the lower part of the ground less stored with herbs. And these so various, that if *Gerard* himself had been in the place, upon the beholding thereof he must have been forced to a re-edition of his *Herbal*, to add the recruit of those *Plants*, which formerly were unseen by him, or unknown unto him.

In this solemn Rendezvous of *Flowers* and *Herbs* the *Rose* stood forth, and made an *Oration* to this effect.

It is not unknown to you, how I have the precedency of all *Flowers* confirmed unto me, under the *Patent* of a double *Sense*, *Sight* and *smell*. What more curious *Colours?* how do all *Dyers* *blush* when they behold my *blushing*, as conscious to themselves that their *Art* cannot imitate that tint which *Nature* hath stamped upon me.

Smell, it is not lusciously *offensive*, nor dangerously *Faint*, but comforteth with a delight, and delighteth with the comfort thereof. Yea, when *Dead*, I am more Sovereign than *Living*. What Cordials are made of my Syrups ? how many corrupted Lungs (those Fans of Nature) sore wasted with consumption, that they seem utterly unable any longer to cool the heat of the *Heart* with their *ventilation*, are, with *Conserves* made of my stamped *Leaves*, restored to their former soundness again. More would I say in mine own cause, but that happily I may be taxed of pride and self-flattery, who speak much in mine own behalf, and therefore I leave the rest to the judgment of such as hear me, and pass from this *discourse* to my just *complaint*.

There is lately a *Flower* (Shall I call it so? In courtesy I will term it so, though it deserve not the appellation), a *Tulip*, which hath ingrafted the love and affections of most people into it; and what is this *Tulip ?* A well-complexioned stink, an ill savour wrapped up in pleasant colours. As for the use thereof in *Physic*, no *Physician* hath honoured it yet with the mention, nor with a *Greek* or *Latin* name, so inconsiderable hath it hitherto been accounted; and yet this is that which filleth all Gardens, hundreds of pounds being given for the root thereof, whilst I, the *Rose*, am neglected and contemned, and conceived beneath the honour of noble hands, and fit only to grow in the gardens of Yeomen. I trust the remainder to your apprehensions, to make out that, which grief for such undeserved injuries will not suffer me to express.

Hereat the *Rose* wept, and the dropping of her *white* tears down her *red* cheeks so well became her, that if ever sorrow was lovely, it then appeared so, which moved the beholders to much compassion, her *Tears* speaking more than her tongue in her own behalf.

The *Tulip* stood up insolently, as rather *challenging*, than *craving* respect, from the *Commonwealth* of *Flowers* there present, and thus vaunted itself.

I am not solicitous what to return to the Complaint of this Rose, whose own demerit hath justly outed itself of that respect which the mistaken world formerly bestowed upon it, and which men's eyes, now opened, justly reassume, and confer on those who better deserve the same. To say that I am not more worthy than the *Rose*, what is it, but to condemn mankind, and to arraign the most *Gentle* and knowing among men, of ignorance for misplacing their affections. Surely *Vegetables* must not presume to mount above *Rational* Creatures, or to think that men are not the most competent judges of the worth and value of *Flowers*. I confess there is yet no known sovereign virtue in my leaves, but it is injurious to infer that I have none, because as yet not taken notice of. If we should examine *all* by their intrinsic value, how many contemptible things in Nature would take the upper hand of those which are most valued. By this argument a *Flint stone* would be better than a *Diamond*, as containing that spark of fire therein, whence men with combustible matter may heat themselves in the coldest season; and clear it is, that the *Loadstone* (that grand *Pilot* to the *North*, which findeth the way there in the darkest night) is to be preferred before the most orient Pearl in the world. But

U

they will generally be condemned for unwise, who prize things according to this proportion.

Seeing therefore in stones and minerals, that those things are not most *valued* which have most virtue, but that men according to their eyes and fancies raise the reputation thereof, let it not be interpreted to my disadvantage, that I am not eminently known for any cordial operation; perchance the discovery hereof is reserved for the next age, to find out the *latent* virtue which *lurketh* in me. And this I am confident of, that Nature would never have hung out so gorgeous a sign, if some guest of quality had not been lodged therein; surely my *leaves* had never been *feathered* with such variety of *colours* (which hath proclaimed me the King of all Lilies) had not some strange virtue, whereof the world is yet ignorant, been treasured up therein.

As for the *Rose*, let her thank herself if she be sensible of any decay in esteem. I have not ambitiously affected superiority above her, nor have I fraudulently endeavoured to supplant her; only I should have been wanting to myself, had I refused those favours from *Ladies* which their importunity hath pressed upon me. And may the *Rose* remember how she, out of causeless jealousy, maketh all hands to be her enemies that gather her; what need is there that she should garrison herself within her prickles? why must she set so many Thorns to lie constant *perdue*, that none must gather her, but such as suddenly surprise her; and do not all that crop her run the hazard of hurting their fingers? This is that which hath weaned the world from her love, whilst my smooth stalk exposing *Ladies* to no such perils, hath made them by exchange to fix their removed affections upon me.

At this stood up the *Violet*, and all prepared themselves with respectful attention, honouring the *Violet* for the Age thereof, for, the *Primrose* alone excepted, it is *Senior* to all the *Flowers* in the year, and was highly regarded for the reputation of the experience thereof, that durst encounter the cold, and had passed many bitter blasts, whereby it had gained much wisdom, and had procured a venerable respect, both to his *person* and *Counsel*.

The case (saith the *Violet*) is not of particular

concernment, but extendeth itself to the life and liberty of all the society of *Flowers*. The complaint of the *Rose*, we must all acknowledge to be just and true, and ever since I could remember, we have paid the *Rose* a just *tribute* of *Fealty*, as our Prime and principal. As for this *Tulip*, it hath not been in being in our *Garden* above these sixty years. Our *Fathers* never knew that such a *Flower* would be, and perhaps our children may never know it ever was; what traveller brought it hither, I know not; they say it is of a *Syrian* extraction, but sure there it grew wild in the open fields, and is not beheld otherwise than a gentler sort of weed. But we may observe that all foreign *vices* are made *virtues* in this country, foreign *drunkenness* is Grecian *Mirth* (thence the proverb, The *Merry Greek*), foreign *pride*, Grecian *good behaviour;* foreign *lust,* Grecian *love;* foreign *laziness,* Grecian *harmlessness;* foreign *weeds,* Grecian *flowers.* My judgment therefore is, that if we do not speedily eradicate this *intruder,* this *Tulip* in process of time will out us all of our just possessions, seeing no *Flower* can pretend a clearer title than the *Rose* hath; and let us every one make the case to be his own.

The gravity of the *Violet* so prevailed with the *Senate* of *Flowers*, that all concurred with his judgment herein: and such who had not the faculty of the fluentness of their tongues to express themselves in large *Orations*, thought that the well managing of a *yea*, or *nay*, spoke them as well wishing to the general good as the expressing themselves in large *Harangues;* and these soberly concluded that the *Tulip* should be rooted out of the Garden, and cast on the *dunghill*, as one who had justly invaded a place not due thereunto; and this accordingly was performed.

Whilst this was passing in the *upper house* of the *Flowers*, no less were the transactions in the lower house of the *herbs;* where there was a general acclamation against *Wormwood*, the generality condemning it, as fitter to grow in a *ditch*, than in a *Garden*. *Wormwood* hardly received leave to make its own defence, pleading in this manner for its innocency.

I would gladly know whom I have offended in this commonwealth of *Herbs*, that there should be so general a conspiracy against me? only two things can be charged on me, *commonness* and *bitterness*. If *commonness* pass for a fault, you may arraign *Nature* itself, and condemn the best Jewels thereof, the *light* of the *Sun*, the *benefit* of the *Air*, the *community* of the *Water;* are not these staple commodities of mankind, without which, no being or subsistence? if therefore it be my charity to stoop so low as to tender myself to every place, for the public service, shall that for which, if I deserve not praise I need no pardon, be charged upon me as an offence?

As for my *bitterness*, it is not a malicious and mischievous *bitterness* to do hurt, but a helpful and medicinal *bitterness*, whereby many cures are effected. How many have surfeited on honey? how many have digged their graves in a Sugar-loaf? how many diseases have been caused by the *dulceor* of many luscious sweetmeats? then am I sent for *Physician* to these patients, and with my brother *Cardus* (whom you behold with a loving eye—I speak not this to endanger him, but to defend myself) I restore them (if temperate in any degree, and persuaded by their friends to taste of us) unto their former health. I say no more; but were all my patients now my pleaders, were all those who have gained *health by me*, present to *intercede for me*, I doubt not but to be reinstated in your good opinion.

True it is, I am condemned for over-hot and too passionate in my operation; but are not the best natures subject to this distemper? is it not observed that the most *witty* are the most *choleric?* a little over-doing is *pardonable*, I will not say necessary in this kind; nor let me be condemned as destructive to the *sight*, having such good opening and *abstergent* qualities, that modestly taken, especially in a *morning*, I am both *food* and *Physic* for a *forenoon*.

It is strange to see how *passion* and *self-interest* sway in many things, more than the justice and merit of a cause. It was verily expected that *Wormwood* should have been acquitted, and re-admitted a *member* in the *Society* of *Herbs*. But what will not a *Faction* carry? *Wormwood's* friends were casually absent that very day, making merry at an entertainment; her enemies (let not that sex be angry for making *Wormwood* feminine) appeared in a full body, and made so great a noise, as if some *mouths* had two *tongues* in them; and though some engaged very zealously in *Wormwood's* defence, yet, over-charged with the *Tyranny* of *Number*, it was carried in the *Negative*, that *Wormwood*, *alias absynthium*, should be plucked up root and branch from the *Garden*, and thrown upon the *Dunghill*, which was done accordingly, where it had the woeful Society of the *Tulip*, in this happy, that being equally miserable, they might be a comfort the one to the other, and spend many hours in mutual recounting their several calamities, thinking each to exceed the other in the relation thereof.

Let us now amidst much sadness interweave something of more mirth and pleasantness in the *Garden*. There were two *Roses* growing upon one *Bush*, the one *pale* and *wan* with *age*, ready to drop off, as useful only for a *Still*; the other a young *Bud*, newly loosened from its *green swaddling-clothes*, and peeping on the rising Sun, it seemed by its orient colour to be dyed by the reflection thereof.

Of these the aged *Rose* thus began. Sister *Bud*,

learn *wit* by my *woe*, and cheaply enjoy the *free* and *full* benefit of that purchase which cost me *dear* and *bitter* experience. Once I was like yourself, young and pretty, straitly laced in my *green Girdle*, not swollen to that breadth and corpulency which now you behold in me, every hand which passed by me courted me, and persons of all sorts were ambitious to gather me. How many fair fingers of curious Ladies tendered themselves to remove me from the place of my abode, but in those days I was coy, and, to tell you plainly, foolish. I stood on mine own defence, summoned my *life-guard* about me, commanded every prickle, as so many *Halberdiers*, to stand to their *Arms*, defy those that touch me, protested myself a votary of constant virginity; frighted hereat, passengers desisted from their intentions to crop me, and left me to enjoy the sullen humour of my own reservedness.

Afterwards the Sun became wrought powerfully upon me (especially about noon-time), to this my present extent, the *Orient colour* which blushed so beautiful in me at the first was much abated, with an over mixture of *wanness* and *paleness* therewith, so that the *Green*, or white sickness rather, the common penance for over-kept Virginity began to infect me, and that fragrant scent of mine began to remit and lessen the sweetness thereof, and I daily decayed in my natural perfume; thus seeing I daily lessened in the repute of all eyes and nostrils, I began too late to repent myself of my former frowardness, and sought that my diligence, by an after-game, should recover what my folly had lost. I pranked up myself to my best advantage, summoned all my sweetness to appear in the height thereof, recruited my decayed *Colour* by blushing for my own folly, and wooed every hand that passed by me to remove me.

I confess in some sort it offers rape to a Maiden modesty, if forgetting their sex, they that should be all *Ears* turn mouths, they that should expect, offer; when we women, who only should be the passive *Counterparts* of *Love*, and receive impression from others, boldly presume to stamp them on others, and by an inverted method of nature, turn pleaders unto men, and woo them for their affections. For all this there is but one excuse, and that is absolute necessity, which as it breaks through stone walls, so no wonder if in this case it alters and transposes the *Sexes*, making women to *man* it in case of extremity, when men are wanting to tender their affections unto them.

All was but in vain; I was entertained with scorn and neglect, the *hardened* hands of *daily Labourers*, *brawned* with continual work, the *black* hands of *Moors*, which always carry *Night* in their *Faces*, slighted and contemned me; yea, now behold my last hope is but to deck and adorn houses, and to be laid as a property in windows, till at last I die in the *Hospital* of some *still*, where, when useless for anything else, we are generally admitted. And now my very *leaves* begin to *leave* me, and I to be deserted and forsaken of myself.

O how happy are those *Roses* who are preferred in their youth to be warm in the hands and breasts of fair Ladies, who are joined together with other *flowers* of *several kinds* in a *Posy*, where the general result of sweetness from them all, ravisheth the *Smell* by an intermixture of various Colours, all united by their *stalks* within the same *thread* that bindeth them together!

Therefore, Sister *Bud*, grow wise by my folly, and know, it is far greater happiness to lose thy *Virginity* in a good hand, than to wither on the *stalk* whereon thou growest: accept of thy just and best tender, lest afterwards in vain thou courtest the reversion of fragments of that feast of love, which first was freely tendered unto thee.

Leave we them in their discourse, and proceed to the relation of the *Tulip* and *Wormwood*, now in a most pitiful condition, as they were lying on the *Dunghill;* behold a *vast Giant Boar* comes unto them; that which *Hercules* was said to kill, and which was accounted by some the *foreman* of the *jury* of his *Labours*, was but a *Pigmy*, or rather but a *pig*, in comparison of this; and with his *Tusks*, wherewith *Nature* had armed him to be his sword, as his *shoulders* are his shield, he began to rend and tear the *Tulip* and *Wormwood*, who exclaimed unto him as followeth :—

Sir, pity useth always to be an attendant of a generous mind and valiant spirit, for which I have heard you much commended. *Cruelty* is commonly observed to keep company with *Cowardliness,* and *base minds* to triumph in *cruel actions;* behold we are the objects rather of your pity, whose sufferings may rather render us to the commiseration of any, that justly consider our case. I, the *Tulip*, by a *Faction* of *flowers*, was ousted of the *Garden*, where I have as good a right and title to abide as any other; and this *Wormwood*, notwithstanding her *just* and long *plea*, how useful and cordial she was, was by a *conspiracy* of *Herbs* excluded the *Garden*, and both of us ignominiously confined to this place, where we must without all hopes quickly expire. Our humble request unto you is not to shorten those few minutes of our lives which are left unto us, seeing such prejudice was done to our *Vitals* (when our roots were mangled by that cruel eradication) that there is an impossibility of our long continuance. Let us therefore fairly breathe out our last breath, and antedate not our misery, but let us have the favour of a quiet close and conclusion.

But if so be that you are affected with the

destruction of *Flowers* and *Herbs*, know the *pleasure* and *contentment* therein must be far greater to root out those which are fairly *flourishing* in their *prime*, whereof plenty are in this *Garden* afforded; and if it please you to follow our directions, we will make you *Master* of a *Pass*, which without any difficulty shall convey you into the *Garden;* for though the same on all sides almost is either *walled*, or *paled* about, yet in one place it is fenced with a *Hedge* only, wherein, through the neglect of the *Gardener* (whose care it ought to be, to secure the same), there is a *hole*, left in such capacity as will yield you an easy entrance thereinto. There may you glut yourself, and satiate your soul with variety of *Flowers* and herbs, so that an *Epicure* might have cause to complain of the plenty thereof.

The *Boar* apprehends the motion, is sensible it was advantageous for him, and following their directions, he makes himself Master of his own desire. O the spitefulness of some *Natures!* how do they *wreak* their anger on all persons. It was revenge for the *Tulip* and *Wormwood*, unless they had spitefully wronged the whole

Corporation of *Flowers*, out of which they were ejected as useless and dangerous *Members*. And now consider, how these two *pride* themselves in their own *vindictive* thoughts; how do they in their forerunning fancy, antedate the death of all *Herbs* and *Flowers!* What is sweeter than revenge? how do they please themselves to see what are *hot* and *cold* in the *first, second, third,* and *fourth degree* (which borders on poison), how all these different in their several *Tempers*, will be made *friends* in universal misery, and *compounded* in a general destruction.

Little did either *Flowers* or *Herbs* think of the *Boar's* approaching, who were solacing themselves with merry and pleasant discourse; and it will not be amiss to deceive time by inserting the *Courtship* of *Thrift*, a *flower-Herb*, unto the *Marigold*, thus accosting her, just as the *Boar* entered into the *Garden*.

Mistress, of all *Flowers* that grow on Earth, give me leave to profess my sincerest affections to you. Compliments have so infected men's tongues (and grown an *Epidemical fault*, or as others esteem it, a fashionable accomplishment) that we know not when they speak truth, having made dissembling their language, by a constant usage thereof; but believe me, *Mistress*, my *heart* never entertained any other interpreter than my *Tongue;* and if there be a *vein* (which Anatomists have generally avouched) carrying intelligence from the *heart* to the *lips*, assure yourself that *vein* acts now in my discourse.

I have taken signal notice of your accomplishments, and among many other rare qualities, particularly of this, your loyalty and faithfulness to the *Sun, Sovereign* to all *Vegetables*, to whose warming *Beams* we owe our *being* and *increase;* such your love thereunto, that you attend hi. *rising* and therewith *open*, and at his setting *shut* your *windows*. True it is, that *Helitropium* (*or turner with the Sun*) hath a long time been attributed to the *Sun-flower*, a voluminous Giant-like *Flower*, of no virtue or worth as yet discovered therein; but we all know the many and Sovereign virtues in your leaves, the *Herb general in all pottage*. Nor do you as Herb *John* stand neuter, and as too many now-a-days in our *Commonwealth*, do neither good nor ill (expecting to be acted on by the impression of the prevalent party), and otherwise warily engage not themselves; but you really appear sovereign and operative in your wholesome effects. The consideration hereof, and no other by reflection, hath moved me to the tender of my affections, which if it be candidly *resented*, as it is sincerely offered, I doubt not but it may conduce to the mutual happiness of us both.

Besides, know (though I am the most improper person to trumpet forth my own praise) my *name* is *Thrift*, and my *nature* answereth thereunto. I do not prodigally waste those Lands in a *moment* which the industry and frugality of my Ancestors hath in a long time advanced. I am no gamester, to shake away with a *quaking* hand what a more *fixed* hand did gain and acquire. I am none of those who in variety of clothes bury my quick estate, as in a winding-sheet; nor am I one of those who by cheats and deceits improve myself on the losses of others; no *Widows* have wept, no *Orphans* have cried for what I have offered unto them (this is not *Thrift* but rather *Felony*), nor owe I anything to my own body. 'I fear not to be arrested upon the *action* of my own carcase, as if my creditors should cunningly compact therewith, and quit scores, resigning their Bill and Bond unto mine own body, whilst that in requital surrendereth all obligations for food and clothes thereunto. Nor do I undertake to buy out *Bonds* in *controversies* for almost nothing, that so running a small hazard, I may gain great advantage, if my bargain therein prove successful. No, I am plain and honest *Thrift*, which none ever did, or will speak against, save such *prodigal spendthrifts*, who in their reduced thoughts will speak more against themselves.

And now it is in your power to accept or refuse what I have offered, which is the *privilege* which nature hath allotted for your *feminine sex*, which we men, perchance, may grudge and repine at; but it being past our power to amend it, we must permit ourselves as well as we may, to the constant custom prevailing herein.

The *Marigold* demurely hung down her head, as not overfond of the motion, and kept silence so long as it might stand with the rule of manners, but at last broke forth into the following return.

I am tempted to have a good opinion of myself, to which all people are prone, and we women most of all, if we may believe your opinions of us, which herein I am afraid are too true. But, Sir, I conceive myself too wise to be deceived by your commendations of me, especially in so large a way, and on so general an account, that other *Flowers* not only share with me, but exceed me therein. May not the *Daisy* not only be *co-rival* with me, but *superior* to me in that quality wherein so much you praise me? My *vigilancy starteth* only from the *Sun rising*, hers bears date from the *dawning* of the *morning*, and outruns my speed by many degrees; my *virtus* in pottage, which you so highly commend, impute it not to my *Modesty*, but to my *Guiltiness*, if I cannot give it entertainment; for how many hundred *Herbs* which you have neglected exceed me therein?

But the plain truth is, you love not me for myself, but for your advantage. It is *Gold* on the *arrear* of my *name* which maketh *Thrift* to be my *Suitor*. How often, and how unworthily have you tendered your affections, even to a *Penny-*

to be arrested upon the *action* of my own carcase, as if my creditors should cunningly compact therewith, and quit scores, resigning their Bill and Bond unto mine own body, whilst that in requital surrendereth all obligations for food and clothes thereunto. Nor do I undertake to buy out *Bonds* in *controversies* for almost nothing, that so running a small hazard, I may gain great advantage, if my bargain therein prove successful. No, I am plain and honest *Thrift*, which none ever did, or will speak against, save such *prodigal spendthrifts*, who in their reduced thoughts will speak more against themselves.

And now it is in your power to accept or refuse what I have offered, which is the *privilege* which nature hath allotted for your *feminine sex*, which we men, perchance, may grudge and repine at; but it being past our power to amend it, we must permit ourselves as well as we may, to the constant custom prevailing herein.

The *Marigold* demurely hung down her head, as not overfond of the motion, and kept silence so long as it might stand with the rule of manners, but at last broke forth into the following return.

I am tempted to have a good opinion of myself, to which all people are prone, and we women most of all, if we may believe your opinions of us, which herein I am afraid are too true. But, Sir, I conceive myself too wise to be deceived by your commendations of me, especially in so large a way, and on so general an account, that other *Flowers* not only share with me, but exceed me therein. May not the *Daisy* not only be *co-rival* with me, but *superior* to me in that quality wherein so much you praise me? My *vigilancy starteth* only from the *Sun rising*, hers bears date from the *dawning* of the *morning*, and outruns my speed by many degrees; my *virtue* in pottage, which you so highly commend, impute it not to my *Modesty*, but to my *Guiltiness*, if I cannot give it entertainment; for how many hundred *Herbs* which you have neglected exceed me therein?

But the plain truth is, you love not me for myself, but for your advantage. It is *Gold* on the *arrear* of my *name* which maketh *Thrift* to be my *Suitor*. How often, and how unworthily have you tendered your affections, even to a *Penny-*

royal itself, had she not scorned to be courted by you.

But I commend the girl that she knew her own worth, though it was but a *penny*, yet it is a *Royal* one, and therefore not a fit match for every base *Suitor*, but knew how to value herself; and give me leave to tell you, that *Matches* founded on *Covetousness* never *succeed*. *Profit* is the *Loadstone* of your *affections*, *Wealth* the *attractive* of your *Love*, *Money* the *mover* of your desire. How many hundreds have engaged themselves on these principles, and afterwards have bemoaned themselves for the same? But oh, the uncertainty of wealth! how unable it is to expiate and satisfy the mind of man. Such as cast Anchor thereat seldom find fast ground, but are *tossed* about with the *Tempests* of many *disturbances*. These *Wives* for *conveniency* of *profit* and *pleasure* (when there hath been no further nor higher intent) have filled all the world with *mischief* and *misery*. Know then, Sir, I return you a ·flat *denial;* a *denial* that *virtually* contains many, yea, as many as ever I shall be able to pronounce. My tongue knows no other language to you but No; score it upon women's dissimulation (whereof we are too guilty, and I at other times as faulty as any). But, Sir, read my eyes, my face, and compound all together, and know these are the expressions dictated from my heart. I shall embrace a thousand deaths sooner than your Marriage Bed.

Thus were they harmlessly discoursing, and feared no ill, when on a sudden they were surprised with the uncouth sight of the *Boar*, which had entered their *Garden*, following his prescribed directions; and armed with the *Corslet* of his *Bristles, vaunted* like a triumphant *Conqueror* round about the *Garden,* as one who would first make them suffer in their *fear* before in their feeling. How did he please himself in the variety of the *fears* of the *flowers!* to see how some *pale* ones looked *red,* and some *red* ones looked *pale ;* leaving it to *Philosophers* to *dispute* and *decide* that different effects should proceed from the same causes. And among all *Philosophers,* commend the question to the *Stoics,* who because they pretend an *Antipathy,* that they themselves would never be angry, never be mounted above the *model* of a common usual *Temper,* are most competent Judges, impartially to give the reason of the causes of the anger of others.

And now it is strange to see the several ways the *Flowers* embraced to provide for their own security. There is no such *teacher* as *extremity*; *necessity* hath found out more *Arts* than ever ingenuity invented. The *Wall Gillyflower* ran up to the top of the Wall of the *Garden*, where it hath grown ever since, and will never descend till it hath good security for its own safety; and being mounted thereon, he entertained the *Boar* with the following discourse.

Thou basest and unworthiest of *four-footed Beasts*: thy *Mother* the *Sow* passeth for the most contemptible *name* that can be fixed on any *She*. Yea, *Pliny* reporteth that a *Sow* grown *old* useth to feed on her own *young*; and herein I believe that *Pliny*, who otherwise might be straitened for *fellow-witnesses*, might find such, who will attest the truth of what he hath spoken. Men's *Excrements* is thy element, and what more cleanly creatures do scorn and detest makes a feast for thee; nothing comes amiss unto thy mouth, and we know the proverb, what can make a pancake unto thee. Now you are gotten into the *Garden* (shame light on that negligent *Gardener*, whose care it was to fence the same, by whose negligence and oversight you have gotten an entrance into this *Academy* of *Flowers* and *Herbs*), let me who am your enemy give you some *Counsel*, and *neglect* it not because it comes from my *Mouth*. You see I am without the reach of your *Anger*, and all your power cannot hurt me except you be pleased to borrow *wings* from some *Bird*, thereby to advantage yourself to reach my habitation.

My *Counsel* therefore to you is this; be not *Proud* because you are *Prosperous*. Who would ever have thought that you could have entered this place, which we conceived was impregnable against any of your kind? Now because you have had success as far above our *expectations* as your *deserts*, show your own moderation in the usage thereof: to *Master* us is easy, to *Master* yourself is difficult. Attempt therefore that, which as it is most *hard* to *perform*, so will it bring most *honour* to you when *executed*; and know, I speak not this in relation to myself (sufficiently privileged from your *Tusks*) but as *acted* with a *public spirit*, for the good of the *Commonalty* of *Flowers*; and if anything hereafter betide you other than you expect, you will remember that I am a *Prophet*, and foretell that which too late you will credit and believe.

The *Boar* heard the words, and entertained them with a *surly silence*, as conceiving himself to be *mounted* above danger. Sometimes he pitied the silliness of the *Wallflower* that pitied him, and sometimes he vowed revenge, concluding that the *stones* of the *Wall* would not afford it sufficient moisture for its constant dwelling there, but that he should take it for an advantage, when it descended for more sustenance.

It is hard to express the *panic* fear in the rest of the *Flowers*; and especially the *small Primroses* begged of their *Mothers* that they might retreat into the middle of them, which would only make them grow bigger and broader; and it would grieve a pitiful heart to hear the child plead, and the mother so often deny.

The Child began: dear *Mother*, she is but half a *Mother* that doth *breed* and not *preserve*.

Only to *bring forth*, and then to expose us to worldly *misery, lessens* your *Love* and *doubles* our *sufferings*. See how this tyrannical *Boar* threatens our instant undoing. I desire only a *Sanctuary* in your *bosom*, a retreating place into your *breast;* and who fitter to come into you than she that came out of you? Whither should we return than from whence we came? it will be but one happiness or one misfortune; together we shall die, or together be preserved; only some content and comfort will be unto me, either to be happy or unhappy in your company.

The broader *Primrose* hearkened unto these words with a sad countenance, as sensible in herself, that had not the present necessity hardened her affections, she neither would nor could return a deaf ear to so equal a motion. But now she rejoined.

Dear Child, none can be more sensible than myself of Motherly affections; it troubles me more for me to deny thee, than for thee to be denied. I love thy safety, where it is not necessarily included in my danger: the entertaining of thee will be my ruin and destruction: how many

Parents in this age have been undone merely for affording house and home to such Children, whose condition might be quarrelled with as exposed to exception.

I am sure of mine own innocency, which never in the least degree have offended this *Boar*, and therefore hope he will not offend me. What wrong and injury you have done him is best known to yourself; stand therefore on your own bottom, maintain your own innocence. For my part I am resolved not to be drowned for others hanging on me; but I will try as long as I can the strength of my own arms and legs. Excuse me, good child; it is not *hatred* to you, but *love* to myself, which makes me to understand my own interest. The younger *Primrose* returned.

Mother, I must again appeal to your affections, despairing to find any other *Judge* to Father my cause. Remember I am part of yourself, and have never by any undutifulness disobliged your affections. I profess also mine own integrity, that I never have offended this *Boar*, being more innocent therein than yourself; for also my tender years entitles me not to any correspondency with him: this is the first minute (and may it be the last) that ever I beheld him. I reassume therefore my suit, supposing that your first denial proceeded only from a desire to try my importunity, and give me occasion to enforce my request with the greater earnestness. By your motherly bowels I conjure you (an exorcism which I believe, comes not within the compass of superstition), that you tender me in this my extremity, whose greatest ambition is to die in those arms from whence I first fetched my original. And then she left her *tears* singly to drop out the remainder, what her *tongue* could not express.

The *Affections* of Parents may sometimes be *smothered*, but seldom *quenched;* and meeting with the *blast* or *bellows* from the submissive mouths of their *Children*, it quickly *blazeth* into a *flame. Mother* and *daughter* are like Tallies, one exactly answereth the other. The *Mother Primrose* could no longer resist the violence of her daughter's importunity, but opens her bosom for the present reception thereof; wherein ever since it hath grown doubled unto this day. And yet a double mischief did arise from this gemination of the *Primrose*, or inserting of the little one into the Bowels thereof.

First, these *Primroses* ever since grow very slowly, and *lag* the last among all *Flowers* of that kind. Single *Primroses* beat them out of distance, and are arrived at their *Mark* a month before the others start out of their *green leaves*. Yet it will not be hard to assign a natural cause thereof; namely, a greater power of the *Sun* is required to the production of greater *Flowers*. Small degrees of heat will suffice to give a being to single *Flowers*, whilst double ones, groaning under the weight of their own greatness, require a greater force of the *Sunbeams* to quicken them, and to spur their *laziness*, to make them appear out of their roots.

But the second *mischief* most concerns us, which is this: all single *Flowers* are *sweeter* than those that are double; and here we could wish that a *Jury* of *Florists* were *impanelled*, not to eat until such time as they were agreed in their *verdict* what is the true cause thereof. Some will say that single *leaves* of *Flowers*, being more effectually wrought on by the *Sunbeams*, are rarefied thereby, and so all their sweetness and perfume the more fully extracted; whereas double *Flowers*, who lie as it were in a lump and heap, crowded together with its own *leaves*, the *Sunbeams* hath not that advantage singly to distil them, and to improve every particular *leaf* to the best advantage of sweetness. This sure I am, that the old *Primrose*, sensible of the abatement of her sweetness, since she was *clogged* with the entertainment of her *Daughter*, half repenting that she had received her, returned this complaining discourse.

· *Daughter*, I am sensible that the *statutes* of *inmates* was founded on very good and solid grounds, that many should not be multiplied within the roof of one and the same house, finding the inconveniency thereof by lodging thee my own *Daughter* within my *Bosom*. I will not speak how much I have lost of my growth, the *Clock* whereof is *set back* a whole month by receiving of you; but that which most grieveth me, I perceive I am much abated in my *sweetness* (the essence of all *Flowers*), and which only distinguisheth them from *weeds*, seeing otherwise, in

Colours weeds may contest with us in brightness and variety.

Please, Mother (replied the small *Primrose*), conceive not this to be your particular unhappiness, which is the general *accident* falling out daily in common experience; namely, that the bigger and thicker *people* grow in their *estates*, the worse and less virtuous they are in their *Conversations:* our age may produce millions of these instances. I know many honest men, whose converse some ten years since was familiar and fair: how did they court and desire the company of their neighbours, and mutually, how was their company desired by them: how *humble* were they in their *carriage*, loving in their *expressions*, and *friendly* in their *behaviour*, drawing the love and affections of all that were acquainted with them! but since being grown wealthy, they have just learnt not to know *themselves*, and afterwards none of their *neighbours*. The *brightness* of much *Gold* and *Silver* hath with the *shine* and *lustre* thereof, so *perstringed* and *dazzled* their *eyes*, that they have forgotten those with whom they had formerly so familiar conversation.

How *proudly* do they *walk*, how *superciliously* do they *look*, how *disdainfully* do they *speak*! They will not know their own *Brothers* and *kindred*, as being akin only to themselves.

Indeed, such who have long been gaining of wealth, and have slowly proceeded by degrees therein, whereby they have learnt to manage their minds, are not so palpably proud as others; but those who in an instant have been surprised with a vast estate, *flowing* in upon them from a *fountain* far above their *deserts*, not being able to wield their own greatness, have been pressed under the weight of their own estates, and have manifested that their minds never knew how to be stewards of their wealth, by forgetting themselves in the disposing thereof.

I believe this little *Primrose* would have been longer in her discourse, had not the approach of the *Boar* put an unexpected period thereunto, and made her break off her speech before the ending thereof.

Now whilst all other *flowers* were struck into a *panic* silence, only *two*, the *Violet* and the *Marigold*, continued their discourse; which was not attributed to their valour or hardiness above other *Flowers*, but that casually both of them grew together in the *declivity* of a depressed *Valley*, so that they saw not the *Boar;* nor were they sensible of their own misery, nor durst others remove their stations to bring them intelligence thereof.

Sister *Marigold*, said the *Violet*, you and I have continued these many days in the contest which of our two *colours* are the most honourable and pleasing to the *Eye*. I know what you can plead for yourself, that your *yellowness* is the *Livery* of *Gold*, the *Sovereign* of most men's hearts, and esteemed the purest of all *metals;* I deny not the truth thereof. But know, that as far as the *Sky* surpasseth that which is buried in the *Bowels* of the *Earth*, so far my blue colour exceedeth yours. What is oftener mentioned by the *Poets* than the *azure Clouds?* Let *Heralds* be made the *Umpires*, and I appeal to *Gerard*, whether the *azure* doth not carry it clear above all other *colours* herein. *Sable* or *Black* affrights the *beholders* with the *hue* thereof, and minds them of the *Funeral* of their last friends whom they had interred. *Vert* or *Green* I confess is a *colour* refreshing the *sight*, and worn commonly before the eyes of such who have had a casual mischance therein: however, it is but the *Livery* of *novelty*, a young upstart *colour;* as *green heads* and *green youth* do pass in common experience. *Red* I confess is a noble *colour*, but it hath too much of *bloodiness* therein, and affrighteth beholders with the memory thereof. My *Blue* is exposed to no cavils and exceptions, wherein *black* and *red* are moderately compounded, so that I participate of the perfections of them both; the over-gaudiness of the *red*, which hath too much *light* and *brightness* therein, is reduced and tempered with such a *mixture* of *black*, that the *red* is made *staid*, but not *sad* therewith, and the *black* kept from over-much melancholy, with a proportionable contemperation of *red* therein. This is the reason that in all ages the *Violet* or *purple colour* hath passed for the emblem of Magistracy, and the *Robes* of the ancient Roman judges always dyed therewith.

The *Violet* scarce arrived at the middle of her discourse, when the approach of the *Boar* put it into a terrible fear; nor was there any *Herb* or *Flower* in the whole *Garden* left unsurprised with fear, save only *Thyme* and *Sage*, which casually grew in an *island* surrounded with water from the rest, and secured with a lock-bridge from the *Boar's* access. *Sage*, beginning, accosted *Thyme* in this Nature.

Most fragrant Sister, there needs no other argument to convince thy transcendent sweetness, save only the appealing to the *Bees* (the most competent judges in this kind), those little *Chemists*, who, through their natural *Alembic*, *distil* the sweetest and usefullest of *Liquors*, did not the commonness and cheapness thereof make it less valued. Now those industrious *Bees*, the emblem of a commonwealth (or monarchy rather, if the received traditions of a *Master-Bee* be true), make their constant diet upon thee; for though no *Flower* comes amiss to their palates, yet are they observed to prefer thee above the rest. Now, Sister *Thyme*, fain would I be satisfied of you several queries, which only *Thyme* is able to resolve. Whether or no do you think that the *State* of the *Turks*, wherein we live (whose cruelty hath destroyed fair *Tempe,* to the small remnant of these few Acres), whether, I say, do you think that their *strength* and *greatness* doth *increase, stand still,* or *abate?* I know, *Thyme,* that *you are the Mother of truth,* and the finder out of all truth's mysteries. Be open therefore, and candid with me herein, and freely speak your mind of the case propounded.

Thyme, very gravely casting down the eyes thereof to the Earth, Sister *Sage* (said she), had you propounded any question within the sphere or circuit of a *Garden,* of the *heat* or *coolness, dryness* or *moisture, virtue* or *operation* of *flowers* and *herbs,* I should not have demurred to return you a speedy answer; but this is of so dangerous consequence, that my own safety locks up my lips, and commands my silence therein. I know your wisdom, *Sage,* whence you have gotten your name and reputation. This is not an age to trust the nearest of our relations with such an important secrecy. Whatever thoughts are concealed within the *Cabinet* of my own *bosom* shall there be preserved in their secret property, without imparting them to any. My confessor himself shall know my *conscience*, but not my judgment in affairs of *State*. Let us comply with the present necessity, and lie at a close posture, knowing there be fencers even now about us who will set upon us if our guards lie open : general discourses are such to which I will confine myself. It is anciently said, *that the subtle man lurks in general*. But now give me leave, for honesty itself, if desiring to be safe, to take Sanctuary therein.

Let us enjoy our own happiness, and be sensible of the favour indulged to us, that whereas all *Tempe* is defaced, this *Garden* still surviveth in some tolerable condition of prosperity; and we especially *miled* about, are fenced from foreign foes better than the rest. Let it satisfy your soul that we peaceably possess this happiness, and I am sorry that the lustre thereof is set forth with so true a foil as the calamity of our neighbours.

Sage returned: Were I a blab of my mouth, whose secrecy was ever suspected, then might you be cautious in communicating your mind unto me. But secrecy is that I can principally boast of, it being the quality for which the commonwealth of *Flowers* chose me their privy Councillor. What therefore is told me in this nature, is deposited as securely as those *treasures* which formerly were laid up in the *Temple* of safety itself; and therefore with all ˙modest importunity, I reassume my suit, and desire your judgment of the question, whether the *Turkish Tyranny* is likely to continue any longer? for *Thyme* I know alone can give an answer to this question.

Being confident (*said Thyme*) of your fidelity, I shall express myself in that freeness unto you which I never as yet expressed to any mortal. I am of that hopeful opinion, that the period of this barbarous nation's greatness begins to approach. My first reason is drawn from the vicissitude and mutability which attends all earthly things. *Bodies* arrived at the *vertical* point of their *strength, decay* and *decline.* The *Moon,* when in the fulness of its *increasing,* tendeth to a *waning.* It is a pitch too high for any sublunary thing to mount unto, constantly to proceed progressively in greatness. This maketh me to hope that this Giant-like *Empire*, cemented with Tyranny, supported, not so much with their own policy, as with the servility of such who are under them, hath seen its best days and highest elevation.

To this end, to come to more particulars, what was it which first made the *Turks* fortunate in so short a time to overrun all *Greece*, but these two things: first, the *dissensions*, second, the *dissoluteness* of your ancient Greeks. Their *dissensions* are too well known, the Emperor of *Constantinople* being grown almost but *titular*, such the pride and potency of many Peers under him. The *Ægean* is not more stored with *Islands* (as I think scarce such a heap or huddle is to be found in all the world again) as *Greece* was with several *factions:* the *Epirots* hated the *Achaians*, the *Mesedans* banded against the *Thracians*, the *Dalmatians* maintained *deadly feud* against the *Wallachians*. Thus was the conquest made easy for the *Turks*, beholden not so much to their own valour as to the *Grecian* discord.

Next to their *dissensions*, their *dissoluteness* did expedite their ruin. Drunkenness was so common among them that it was a sin to be sober; so that I may say all *Greece reeled* and staggered with its own *intemperance* when the *Turk* assaulted it. What wonder then was it, if they so quickly overran that famous *Empire*, where *vice* and *laziness* had generally infected all conditions of people.

But now you see the *Turks* themselves have divisions and dissensions among them. Their *great Bashaws* and *holy Muftis* have their several factions and dissensions; and whereas the *poor Greeks*, by the reason of their hard usage, begin now to be starved into unity and temperance, they may seem to have changed their vices with the *Turks*, who are now grown as factious and vicious as the others were before. Add to this that they are universally hated, and the neighbouring *Princes* rather *wait a time* than *want a will* to be revenged on them for their many insolences. Put all these together, and tell me if it put not a cheerful complexion on probability, that the *Turkish tyranny*, having come to the mark of its own might, and utmost limits of its own greatness, will dwindle and wither away by degrees. And assure yourself, if once it come to be but *standing water*, it will quickly be a *low ebb* with them.

Probably she had proceeded longer in her Oration, if not interrupted with the miserable moans and complaints of the *Herbs* and *Flowers* which the Boar was ready to devour; when presently the *Sage* spake unto the *Boar* in this manner.

Sir: listen a little unto me, who shall make such a motion, whereof yourself shall be the *Judge* how much it tendeth to your advantage, and the deafest ears will listen to their own interest. I have no design for myself (whose position here, environed with water, secureth me from your anger), but I confess I sympathise with the misery of my friends and acquaintance which in the continent of the *Garden* are exposed to your cruelty. What good will it do you, to destroy so many *Flowers* and *Herbs*, which have no gust or sweetness at all in them for your palate? Follow my directions, and directly *Southwest*, as you stand, you shall find (going forward therein) a corner in the *Garden* overgrown with *Hogweed* (through the *Gardener's* negligence).

THE SPEECH OF FLOWERS.

Oh what *Lettuce* will be for your *lips:* you will say that *Via lactea* (or the milky way) is truly there, so white, so sweet, so plentiful a liquor is to be distilled out of the leaves thereof, which hath gotten the name of *Hogweed,* because it is the principal *Bill of Fare* whereon creatures of your kind make their common repast.

The *Boar*, sensible that *Sage* spake to the purpose, followed his direction, and found the same true; when feeding himself almost to surfeit on those delicious dainties, he swelled so great, that in his return out of the *Garden,* the hole in the fence which gave him *admittance,* was too small to afford him *egress* thereat; when the *Gardener,* coming in with a *Guard* of *Dogs,* so persecuted this *Tyrant,* that, killed on the place, he made satisfaction for the wrong he had done, and for the terror wherewith he had affrighted so many *Innocents*.

I wish the *Reader* well *feasted* with some of his *Brawn,* well cooked, and so take our leave both of him and the *Gardens.*

FINIS.

LONDON:
PRINTED BY W. CLOWES AND SONS, STAMFORD STREET
AND CHARING CROSS.

www.ingramcontent.com/pod-product-compliance
Lightning Source LLC
Chambersburg PA
CBHW030749230426
43667CB00007B/901